Audel™

Machine Shop Tools and Operations
All New 5th Edition

Rex Miller
Mark Richard Miller

WILEY

Wiley Publishing, Inc.

Vice President and Executive Group Publisher: Richard Swadley
Vice President and Executive Publisher: Robert Ipsen
Vice President and Publisher: Joseph B. Wikert
Executive Editorial Director: Mary Bednarek
Editorial Manager: Kathryn A. Malm
Executive Editor: Carol A. Long
Senior Production Editor: Fred Bernardi
Development Editor: Kevin Shafer
Production Editor: Pamela M. Hanley
Text Design & Composition: TechBooks, Inc.

Library of Congress Cataloging-in-Publication Data:

ISBN: 0-764-55527-8

Printed in the United States of America

10 9 8 7 6 5 4 3 2 1

Contents

Acknowledgments

A number of companies have been responsible for furnishing illustrative materials and procedures used in this book. At this time, the authors and publisher would like to thank them for their contributions. Some of the drawings and photographs have been furnished by the authors. Any illustration furnished by a company is duly noted in the caption.

The authors would like to thank everyone involved for his or her contributions. Some of the firms that supplied technical information and illustrations are listed below:

Alcoa

Aluminum Co. of America

Armstrong-Blum Mfg. Co.

Atlas Press Co.

Buffalo Forge

Bullard Co.

Cincinnati Milacron Co.

Ex-Cell-O Corp.

Fred V. Fowler Co.

Giddings and Lewis Machine Tool Co.

Gisholt Machine Co.

Gulton Industries

Hardinge Brothers, Inc.

Heald Machine Co.

Henry G. Thompson Co.

Jacobs Manufacturing Co.

Kalamazoo Saw Div. KTS Industries

Lufkin Rule Co.

Machinery's Handbook, The Industrial Press

Monarch Cortland

Morse Twist Drill and Machine Co.

National Drill and Tool Co.

Nicholson File Company

Norton

Prab Robots
Raytheon
Rockwell Manufacturing
Royal Products
South Bend Lathe, Inc.
Startrite, Inc.
Ultimation
Wilton Tool Manufacturing Co., Inc.

About the Authors

Rex Miller was a Professor of Industrial Technology at The State University of New York—College at Buffalo for over 35 years. He has taught on the technical school, high school, and college level for well over 40 years. He is the author or coauthor of over 100 textbooks ranging from electronics to carpentry and sheet metal work. He has contributed more than 50 magazine articles over the years to technical publications. He is also the author of seven Civil War regimental histories.

Mark Richard Miller finished his B.S. degree in New York and moved on to Ball State University where he obtained the master's and went to work in San Antonio. He taught in high school and went to graduate school in College Station, Texas, finishing the doctorate. He took a position at Texas A&M University in Kingsville, Texas, where he now teaches in the Industrial Technology Department as a Professor and Department Chairman. He has coauthored seven books and contributed many articles to technical magazines. His hobbies include refinishing a 1970 Plymouth Super Bird and a 1971 Roadrunner. He is also interested in playing guitar, which he did while in college as lead in The Rude Boys band.

Introduction

The purpose of this book is to provide a better understanding of the fundamental principles of machine shop practices and operations for those persons desiring to become machinists. The beginning student or machine operator cannot make adequate progress in his or her trade until he or she possesses a knowledge of the basic principles involved in the proper use of various tools and machines commonly found in the machine shop.

One of the chief objectives has been to make the book clear and understandable to both students and workers. The illustrations have been selected to present the how-to-do-it phase of many of the machine shop operations. The material presented here should be helpful to the machine shop instructor, as well as to the individual student or worker who desires to improve himself or herself in this trade.

The proper use of machines and the safety rules for using them have been stressed throughout the book. Basic principles of setting the cutting tools and cutters are dealt with thoroughly, and recommended methods of mounting the work in the machines are profusely illustrated. The role of numerically controlled machines is covered in detail with emphasis upon the various types of machine shop operations that can be performed by them.

This book is presented at a time when there is a definite trend toward expanded opportunities for vocational training of young adults. An individual who is ambitious enough to want to perfect himself or herself in the machinist trade will find the material presented in an easy-to-understand manner, whether studying alone or as an apprentice working under close supervision on the job.

Chapter 1

Power Hacksaws, Power Band Saws, and Circular Saws

Power hacksaws, power band saws, and circular saws are very important to machine shop operations. A large number of power hacksaws and power band saws are in use in the metalworking industry.

Power Hacksaws

A power hacksaw is an essential machine in most machine shop operations. For many years a hand-operated hacksaw was the only means for sawing off metal. Power-driven machines for driving metal-cutting saw blades have been developed to make the task easier. The power hacksaw can do the work much more rapidly and accurately. The machinist should be familiar with these machines, the blades used on the machines, and the operations performed on them. One type of power hacksaw commonly found in machine shops is shown in Figure 1-1.

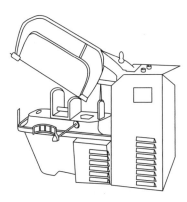

Figure 1-1   Power hacksaw.

Basic Construction

Power hacksaws are designed to make the sawing of metal a mechanical operation. The stock is usually held in a vise mounted on the base of the machine. An electric motor is used to supply power for the machine.

Drive Mechanism
The drive shaft is connected by a V-belt and gears to the electric motor mounted on the machine. The drive mechanism is shielded by guards for safety in operation.

Frame
A U-shaped frame is used on the smaller power hacksaws to support the two ends of the saw blade, which is under tension. The heavier machines use a four-sided frame and a thin backing plate for the blade of the saw.

Worktable and Vise
Most worktables are equipped with a vise that can be mounted either straight or angular to the blade. The worktable is usually mounted on a ruggedly constructed base. Many worktables are provided with T-slots for the purpose of supporting special clamping devices.

Special Features
Nearly all power hacksaws raise the blade on the return stroke. This feature prevents dulling of the blade by dragging it over the work as the blade is returned to the starting position.

Another important feature is a blade safety switch that automatically stops the machine if the blade should break during operation of the saw. The safety switch prevents any damage that could result if the machine continued operation with a broken blade.

Coolant System
Some power hacksaws are equipped with a coolant system that delivers a coolant to the hacksaw blade. The coolant passes from a receiving tank to a pump and then to the work. The machine is equipped with a trough to catch the coolant, which may be screened to remove any chips of metal.

Saw Capacity
Small power hacksaws can be used on square or round stock ranging from ⅛ inch to 3 inches. The larger machines have a capacity ranging to 12 inches (square or round), or even larger.

The capacity of a machine for angular cuts is different from its capacity for straight cuts. The cutting surface is longer for angular cutting. Thus, the saw must be equipped not only with a swivel vise but also with a long enough stroke to make the angular cut.

Blades
High-speed tungsten steel and high-speed molybdenum steel are the most commonly used materials in power saw blades. If only the teeth are hardened, the blades are called *flexible blades*.

Power hacksaw blades are ordered by specifying length and width, thickness, and teeth per inch. For example, they are available in 12″ × 1″ length and width at 0.050-inch thickness. The teeth per inch (TPI) would be either 10 or 14. The 14″ × 1″ × 0.050″ blade is available in only 14 TPI. Blades also come in 17″ × 1″ with a thickness of 0.050 inch and either 10 or 14 TPI. Blades with a thickness of 62 thousandths of an inch (0.062 inch) are usually 1¼ inches wide.

High-speed, shatterproof blades are designed to meet safety and performance requirements. The high-speed molybdenum blades are longer wearing and give the best results for general use. They, too, come in 12-inch, 14-inch, and 17-inch lengths with a 1-inch or 1¼-inch width. These blades are made in both 10 and 14 TPI sizes. Thickness of the metal being cut determines the number of teeth per inch chosen to do the job. There should be no fewer than two or three teeth touching the metal being cut.

Figure 1-2 shows the power hacksaw blade end with a hole for mounting in the machine and the pointed nature of the teeth.

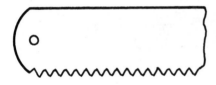

Figure 1-2   Power hacksaw blade.

Hacksaw Operation

Straight cuts are made easily on power hacksaws. The vise is stationary, and the cut is made at a right angle to the sides of the stock.

Most machines are equipped with an adjustable vise. Angular cuts at any desired angle up to 45 degrees can be made by swiveling the vise.

To operate the power saw properly, the work should be fastened securely in the vise so that the blade will saw in the proper place. The blade will break if the work loosens in the vise.

The saw blade should be lowered onto the work carefully to start the cut. On some machines this is done by hand, but it can be done automatically on some saws. In either method, the points of the teeth will be broken or damaged if the blade is permitted to strike the work suddenly.

The machine should be watched carefully to make certain that the saw blade lifts about ⅛ inch on the return stroke. If the blade fails to lift, adjustments should be made immediately, as the blade will be damaged if operation is continued.

When making angular cuts with the work turned at an angle in the vise, another precaution is to be certain that the saw blade can make both the backward and the forward strokes without the saw frame making contact with either the work or the vise. Serious damage to the machine can result from failure to observe this precaution.

When a saw blade is replaced, or a new blade is started in an old cut, it should be remembered that the set of the new blade is wider. The new blade will stick in the old cut unless the work is rotated in the vise a quarter turn. If the work cannot be rotated, the new blade should be guided into the old cut.

The cutting speed of a power hacksaw, of course, varies with the material being cut. Suggested cutting speeds are as follows: for mild steel, 130; for tool steel (annealed), 90; and for tool steel (unannealed), 60. For example, on a machine with a 6-inch stroke, the revolutions per minute of the driving crank should be 130.

All steels should be cut with a cutting compound. Bronze should be cut with a suitable compound at the same speed as mild steel. The saw blade will heat rapidly if an attempt is made to cut brass without a cutting compound adapted to brass. Brass may be cut at the same speed as steel if a suitable compound is used.

Power Band Saws

In the past few years the power band saw has become very important in machine shop operations. In some instances it is used in production operations prior to final machining operations.

Basic Construction

Power band saws are also designed to make the sawing of metal a mechanical operation (Figure 1-3). The stock can be held in a vise mounted on the machine or it can be supported by the operator's hand. Electric motors are used to supply the power for the band saws.

Drive Mechanism

Wheels on the power band saw can be adjusted to apply tension to the band saw blade, which is a flexible, thin, narrow ribbon of steel. One of these wheels is powered by the electric motor mounted on the machine. These wheels are enclosed by guards for safety in operation.

Frame

Many variations of power band saws are available. However, band saws can be grouped into three classifications: horizontal machines for cut-off sawing, vertical machines for straight and profile sawing

Figure 1-3   **A power band saw.** *(Courtesy Kalamazoo Saw Div., KTS Industries.)*

at conventional speeds, and vertical machines for nonferrous cutting and friction cutting.

Worktable and Vise
Power band saws have either a worktable or vise to hold the metal that is to be cut. The worktable is usually part of the band saw. The vise, if there is one, is usually designed to fit the worktable. Many worktables also have T-slots for the purpose of supporting special clamping devices to hold the work to be cut.

Special Features
Some power band saws have automatic tensioning devices so that the proper amount of tension is applied to the band saw at all times. This type of device reduces excessive wear and damage to the blade.

Coolant System
Coolant systems are common on power band saws designed for high-speed production work. These systems deliver coolant to the

band saw blade and the work. After the coolant has been used, it is recycled through a screen and filter to remove any chips.

Capacity of Power Band Saws

The maximum capacity of power band saws varies according to the size of the band saw. Some of the larger machines can accommodate work that is 18″ × 18″ and larger. Band saws also have the capacity to do a wide variety of operations that include cut-off, straight sawing, and contour, or profile, sawing. Rough shaping and semifinishing can be accomplished on almost all types of ferrous and nonferrous materials.

Blades

The development of the heavy-duty band saw machine tool has been designed and built to operate with high-speed blades at maximum efficiency. Use of the band saw offers several important advantages over the hacksaw. These advantages include a narrower kerf, which results in greatly reduced losses; fast, efficient cutting with low per-cut costs; and consistently smooth, accurate cutting.

(A) Standard or regular tooth.

(B) Skip tooth.

(C) Hook tooth.

Figure I-4   Three tooth styles for blades.

The type of blade used on the power band saw is very important. Blades are available with many different types of teeth, tooth set, and pitch.

The three types of teeth are the regular tooth, skip tooth, and hook tooth (Figure 1-4). The regular tooth blade provides from 3 to 32 teeth per square inch. Teeth may be raker set or wavy set. Regular tooth blades are preferred for all ferrous metals and for general-purpose cutting.

Skip-tooth blades feature widely spaced teeth (usually from 2 to 6 TPI) to provide the added chip clearance needed for cutting softer materials. Teeth of the skip-tooth blade are characterized by straight 90° faces and by the sharp angles at the junction of the tooth and gullet. Since these sharp angles and flat surfaces tend to break up chips, this type of tooth is preferred for the very soft nonferrous metals that would otherwise tend to clog and gum the blade. This blade is also widely used in woodcutting.

Hook-tooth blades provide the same wide tooth spacing as the skip-tooth blade. The teeth themselves have a 10° undercut face. Gullets are deeper with blended radii between the teeth and the gullets. Since the undercut face helps the teeth dig in and take a good cut, the blended radii tend to curl the chips. This type of tooth is preferred for the harder nonferrous alloys and many plastic operations.

There are three basic types of tooth set on metal-cutting band saw blades. Each tooth set is designed for a specific type of cutting application. These three basic types of tooth sets are the raker set, the wavy set, and the alternate set. See Figure 1-5.

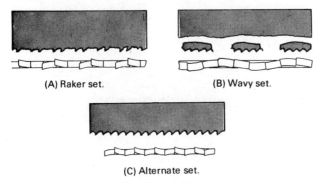

(A) Raker set.   (B) Wavy set.

(C) Alternate set.

Figure 1-5   Types of tooth sets for blades.

The raker set—one tooth set left, one tooth set right, and one tooth not set—is preferred for all long cutting runs where the type of material and the size and shape of the work remain relatively constant. It is also for all contour, or profile, cutting applications and for band sawing with high-speed steel blades.

The wavy-set blade, a rolled set with alternate left and right waves, provides greater chip clearance and a stronger, nearly rip-proof tooth. This blade is preferred for general-purpose work on conventional horizontal machines.

The alternate set has every tooth set—one to the left, one to the right—throughout the blade. Taken originally from the carpenter handsaw, it was supplied in band saws for brass foundry applications.

Band Saw Operation

When operating a band saw, there are some general principles that should be followed so that you will be able to select the blade with the proper pitch:

- Small and thin-wall sections of metal require fine teeth.
- Large metal sections require the use of blades with coarse teeth so that adequate chip clearance is provided.

- Two teeth should be engaged in the metal to be cut at all times.
- Soft, easily machined metals require slightly coarser teeth to provide chip clearance. Hard metals of low machinability require finer teeth so that there are more cutting edges per inch.
- Stainless steel should be sawed with a coarse (NQ) tooth saw blade for best results.

Figure 1-6 shows tooth pitches.

RIGHT   WRONG   RIGHT   LARGE SECTION

(A) Correct and incorrect tooth pitches.

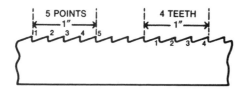

(B) How determined.

Figure 1-6   Tooth pitches.

Select the widest blade that your machine will accommodate when straight, accurate cutting is required. Blade widths for contour sawing depend upon the specific application.

Profile (or contour) sawing is an accurate, fast, and efficient method of producing complex contours in almost any machinable metal. With the proper blade selection, radii as small as $\frac{1}{16}$ inch can be cut. Either internal or external contours can be sawed. If an internal contour has to be sawed, it is first necessary to drill a hole within the contour area to accommodate the saw blade. The band saw blade has to be cut to length, threaded through the drilled hole, and then rewelded. It is for this reason that band saws designed and built for contour sawing have a built-in butt welder. Blades for profile sawing are always raker set because this type of tooth provides

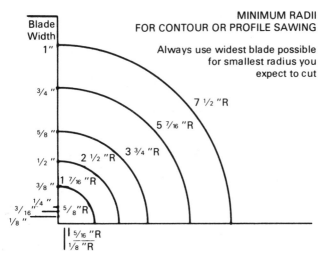

Figure 1-7 Bandwidth selection for profile or contour sawing.

the necessary side clearance. These blades are usually narrower than those chosen for straight cutting; the extra width for any specific job depends upon the smallest radius to be cut. Figure 1-7 will aid you in selecting the correct blade for any application.

Profile sawing requires cutting and rewelding the blade. Because of this, it is necessary to make a good weld. This can be accomplished very easily if certain steps are followed. The steps are as follows:

  1. Square both ends of the blade accurately.
  2. Align both ends of the blade so that they join evenly across the entire width of the blade.
  3. Select the correct dial setting for the blade width.
  4. Select the correct annealing period.
  5. Dress the blade to remove the welding flash.
  6. Anneal for stress relieving.

With the exception of cast iron, coolant should be used for all band-sawing operations. Cast iron is always cut dry. Almost all of the commercially available soluble oils or light cutting oils will give good results when cutting ferrous metals. When aluminum is cut, paraffin and beeswax are commonly used lubricants.

The recommendations shown in Table 1-1 are based on four types of cutting fluids:

A. *Cutting oil mineral base with fatty oils added* (has sulfur for antiweld characteristics and chlorine for film strength)

B. *Cutting oil with light viscosity fatty oils* (for cold-weather operations, mix with (A); when cutting copper alloys, additives and chlorine should be added)

C. *Synthetic water-soluble cutting agent*

D. *Soluble oil cutting agent with fatty oils* (sulfurized for extreme pressure with antiweldment properties; water-soluble for greater heat removal)

Table 1-1   Recommended Cutting Fluids

Materials to Be Cut—Cutting Fluids Recommended

Operation	Free-Cutting Steels; Carbon Steels; Cast Steels; Malleable Iron	Alloy Steels; Nickel Steel; Nickel-Chromium Steel	Stainless Steels (Free Cutting); Ingot Iron; Wrought Iron	Stainless (Austenitic); Manganese Steels; High-Temp Alloys	Aluminum Alloys; Leaded Brass; Magnesium Alloys; Phosphor Bronze; Zinc	Brass; Gun Metal; Nickel; Monel; Inconel
Production sawing using MARVEL high-speed steel, high-speed-edge, and intermediate alloy steel band blades	FLOOD	FLOOD	FLOOD	FLOOD	FLOOD	FLOOD
	(B)	(A)	(A)	(A)	(B)	(A)
	(C) Mix 1 to 5 with water.	(C) Mix 1 to 5 with water.	(D) Mix 1 to 5 with water.	(C) Mix 1 to 5 with water.		(C) Mix 1 to 5 with water.
	(D) Mix 1 to 5 with water.	(D) Mix 1 to 5 with water.	(C) Mix 1 to 5 with water.	(D) Mix 1 to 5 with water.		(D) Mix 1 to 5 with water.

Table 1-1   *(continued)*

Materials to Be Cut—Cutting Fluids Recommended

Operation	Free-Cutting Steels; Carbon Steels; Cast Steels; Malleable Iron	Alloy Steels; Nickel Steel; Nickel-Chromium Steel	Stainless Steels (Free Cutting); Ingot Iron; Wrought Iron	Stainless (Austenitic); Manganese Steels; High-Temp Alloys	Aluminum Alloys; Leaded Brass; Magnesium Alloys; Phosphor Bronze; Zinc	Brass; Gun Metal; Nickel; Monel; Inconel
General sawing using MARVEL hard-back & flexible-back hardened-edge band blades	DRIP (blade just wet)	DRIP (blade just wet)	DRIP (blade just wet)	DRIP (blade just wet)	DRIP (blade just wet)	DRIP (blade just wet)
	(B)	(B)	(B)	(A)	(B)	(A)
	(D)	(D)	(D)	(B)	Paraffin dry	(B)
Contour sawing using MARVEL high-speed-edge or hard-back & flexible-back carbon blades	SPRAY OR DRIP	SPRAY OR DRIP	SPRAY OR DRIP	FLOOD OR DRIP	FLOOD	
	(C)	(C)	(C)	(D) Mix 1 to 10 with water.	(B)	(B)
	(D) Mix 1 to 10 with water.	(D) Mix 1 to 10 with water	(D) Mix 1 to 10 with water. (B)	(C) Mix 1 to 10 with water. (A)	Paraffin dry	(A)

Courtesy Armstrong-Blum Mfg. Co.

The three types of feeds used on power band saws are hand, hydraulic, and mechanical. The number of factors involved in the efficient and trouble-free operation of a power band saw make it difficult to provide specific information on the proper speed or pressure required (see Table 1-2). When a cut is started, the blade should not be forced. If the work is forced into the blade, it will result in a shorter blade life and defective work. The following principles should be used as a guide in selecting the proper feed pressure:

- Use moderate feed pressure when straight, accurate cutting is desired.
- Avoid heavy feed pressures because they will cause the power band saw to chatter and vibrate.
- Study the chips produced. Fine powdery chips indicate the feed is too light. The teeth on the blade are rubbing over the surface of the work instead of cutting the work. Discolored, blued, or straw-colored chips indicate that there is too much feed pressure. This excess pressure can cause teeth to chip and break. In addition, the blade will wear prematurely because of overheating. A free-cut curl indicates ideal feed pressure with the fastest time and the longest blade life.

Table 1-2   Common Band Sawing Problems and Their Connections

Problem	Cause	Correction
Blade develops camber.	Roller guides not correctly adjusted.	Readjust.
	Feeding pressure is too heavy.	Use lighter feed.
	Saw guides too far apart.	Adjust closer to work.
	Blade riding against flange on metal wheel.	Tilt wheel for proper blade position.
Blade develops twist.	Saw is binding in cut.	Decrease feeding pressure.
	Side inserts or rollers of saw guides too close to saw.	Readjust.
	Wrong width of blade for radius being cut.	Check for correct blade width.

Table 1-2   (continued)

Problem	Cause	Correction
Saw dulls prematurely.	Saw speed too great; teeth sliding over work instead of cutting.	Reduce saw speed.
	Improper coolant or coolant improperly directed.	Check type and mixture of coolant. Apply at the point of cut, saturating teeth evenly.
	Saw idling through cut.	Keep teeth engaged; use positive feeding pressure.
	Feed too light; teeth sliding over work.	Use a feed heavy enough to generate a full curled chip.
Saw loses set prematurely.	Saw is too wide for radius being cut.	Check for the proper width.
	Saw speed is too fast.	Reduce speed.
	Saw rubbing against vise or running deep in guides.	Check blade along complete travel.
Saw vibrates in cut.	Wrong speed for the material and work thickness.	Check for the proper width.
	Insufficient blade tension.	Increase tension.
	Pitch too coarse.	Select a finer pitch.
	Excessive feeding pressure.	Reduce pressure.
Saw teeth rip out.	Pitch too coarse.	Use a finer pitch on thin work sections.
	Work not tightly clamped or held.	Reclamp or hold more firmly to prevent vibration.
	Sawing dry.	Use coolant when possible.
	Gullets loading.	Use a coarser pitch and/or a higher viscosity lubricant— or brush to remove chips.
	Excessive feed pressures.	Reduce.

(continued)

Table 1-2 (continued)

Problem	Cause	Correction
Saw breaks prematurely.	Blade too thick for the diameter of wheels.	Check recommendations for your machine.
	Cracking at the weld.	Weld improperly made; try a longer annealing period.
	Pitch is too coarse.	Use finer pitch.
	Excessive feeding pressure or blade tension.	Reduce.
	Guides too tight.	Readjust.

Courtesy Henry G. Thompson Co., Subsidiary of Vermont American Corporation.

One of the most important factors in successful band sawing is a proper cutting speed. If the machine is operated at too fast a speed for the material being cut, the teeth are not allowed sufficient time to dig into the material. As a result, they merely rub over the surface of the work, creating friction that rapidly dulls the cutting edge and wears out the blade. Table 1-3 gives average speeds and cutting rates for cutting a wide variety of commonly used ferrous and nonferrous metals and nonmetallics. These recommendations will ensure optimum performance under most conditions.

Table 1-3 Average Speeds and Cutting Rates

	Flexible and Hard-Back Carbon—Raker Tooth							
	Size of Material							
	¼ in.		**½ in.**		**2 in.**		**4 in.**	
Material	**Speed**	**Teeth**	**Speed**	**Teeth**	**Speed**	**Teeth**	**Speed**	**Teeth**
Steels:								
Armor plate	150	18	125	12	75	8	50	6
Angle iron	175	24	150	14				
Carbon steels	250	24	200	14	150	10	100	6
Chromium steels	150	24	125	14	100	10	50	6
Cold-rolled steel	250	12	200	10	150	8	125	6
Drill rod	100	14	100	14				
Graphite steels	175	18	150	14	100	10	75	6
High-speed steels	150	24	100	14	75	10	50	8

Table 1-3 (continued)

Flexible and Hard-Back Carbon—Raker Tooth

Material	Size of Material							
	1/4 in.		1/2 in.		2 in.		4 in.	
	Speed	Teeth	Speed	Teeth	Speed	Teeth	Speed	Teeth
Machinery steels	250	18	200	14	150	10	125	6
Molybdenum steels	150	18	125	14	75	10	50	8
Nickel steels	150	18	125	14	75,	10	50	8
Silicon manganese	175	18	150	14	75	10	50	6
Stainless steels	100	24	100	14	50	10	50	6
Structural steels	175	24	150	14				
Tungsten steels	175	18	150	14	75	8	50	6
Foundry metals:								
Brass—hard	500	18	400	14	300	10	200	6
Brass—soft	1500	18	1000	14	750	10	300	6
Bronze—aluminum	500	18	400	14	225	10	100	6
Bronze—manganese	300	18	250	14	200	10	150	6
Bronze—navel	300	18	275	14	225	10	150	6
Bronze—phosphorous	500	18	400	14	200	10	150	6
Cast iron—gray	200	18	175	14	100	8	75	6
Cast iron—malleable	200	18	175	14	150	10	125	6
Cast steel	225	18	200	14	100	8	75	6
Copper—beryllium	400	18	350	12	250	10	150	6
Copper—drawn	1100	18	700	10	350	6	200	6
Gunnite	300	24	200	18	150	10	100	6
Meehanite	150	18	100	14	75	8	50	6
Monel	200	18	150	14	75	10	50	6
Nickel—cold-rolled	200	14	150	10	75	8	50	6
Nickel silver	250	18	250	14	175	10	125	6
Silver	250	24	250	18	175	10	150	6

(continued)

Table 1-3 (continued)

Flexible and Hard-Back Carbon—Hook Tooth

	Size of Material							
	1/2 in.		2 in.		5 in.		10 in.	
Material	Speed	Teeth	Speed	Teeth	Speed	Teeth	Speed	Teeth
Nonferrous metals:								
Aluminum—soft	4000	4	3600	3	3200	3	3000	2
Aluminum—medium	3000	6	2000	4	1500	3	1200	2
Aluminum—hard	600	6	500	4	400	3	300	3
Babbit	4000	6	3500	4	3000	3	2500	3
Beryllium copper	2000	6	1800	4	1400	3	1000	3
Brass—casting	3200	6	3000	4	2500	3	1800	3
Brass—commercial	3700	4	3500	3	3000	3	2500	2
Brass—naval	3500	4	3000	3	2500	3	2000	2
Brass—yellow	3200	4	3000	3	2500	3	1800	2
Cadmium—Kirsite	3200	6	3000	4	2500	3	1700	3
Copper	3000	4	3000	3	2500	3	2000	2
Lead	4000	6	3500	4	2700	3	2000	3
Magnesium	4000	6	3000	4	2500	3	2000	2
Silicon bronze	1000	6	1000	4	600	3	300	3
Titanium #150A	60	6	50	6	50	4	50	3
Steels:								
Alloy steels	80	6	70	6	60	4	50	3
Low-carbon steel	170	6	150	4	125	3	100	2
Medium-carbon steel	120	6	110	4	100	3	90	2
High-carbon steel	90	6	80	6	70	4	65	3

Courtesy Armstrong-Blum Mfg. Co.

Friction cutting differs from all other types of metal-sawing methods. It is not actually a cutting operation, but a burning process similar to torch cutting. It is much faster than conventional sawing methods. Very hard materials (which could not normally be sawed) are cut rapidly and easily. Bulky and irregular shapes can be cut handily because there is little blade drag.

Basically, the friction-cutting operation involves the use of a very fast moving blade that travels at speeds between 6000 and 18,000 feet per minute (see Table 1-4). At these speeds, terrific heat is built up in the workpiece at its point of contact with the blade, and burning begins. Teeth are not needed to generate burning heat, but they generate increased cutting speed and efficiency by carrying additional oxygen into the work area, thus creating greater oxidation. Set teeth are also important. They give the operator the control required to cut a straight line or follow a specific curve.

Table 1-4   Recommended Friction Cutting Speeds for Common Metals

Description	Average Recommended Cutting Speeds, ft/min.
Armor plate	7000–13,000
Carbon steel	6000–12,000
Cast steel	7000–13,000
Chromium steel	8000–15,000
Chromium-vanadium steel	8000–15,000
Free-machining steel	6000–12,000
Gray cast iron	7000–13,000
Malleable cast iron	7000–13,000
Manganese steel	6000–12,000
Moly steel	8000–15,000
Molybdenum	8000–15,000
Nickel-chromium steel	8000–15,000
Nickel steel	8000–15,000
Silicon steel	8000–15,000
Stainless steel	7000–13,000
Tungsten steel	8000–15,000

Courtesy Henry G. Thompson Co., Subsidiary of Vermont American Corporation.

There is a definite advantage to using a special blade for this type of operation. The friction-cutting blade is made from special steel. The blade is specially heat-treated to withstand the fast speeds and severe flexing that are encountered in friction sawing.

Many difficult manufacturing problems have been solved with friction sawing because this process has distinct advantages. Before deciding whether to use friction sawing, consideration should be given to the specific job requirements. There are important limitations involved with friction sawing. Following are some of these limitations:

- Friction cutting leaves a very heavy burr on the underside of the material being cut.
- Thicknesses over $5/8$ inch are extremely difficult to cut.
- Heat from the friction-sawing process may have a tempering effect on the edges of some materials.

Circular Saws

Circular saws have some advantages. They produce a burr-free mill finish when slow-speed cold sawing is used. This eliminates secondary operations on tubing, channels, angles, and solid stock of most steels and other ferrous materials, as well as most nonferrous metals. The rigidity of the blade produces cuts of extreme accuracy and close tolerances. The cutting operation is safe, clean, and quiet because of the slower speed.

Figure 1-8 shows a manually operated saw. Semiautomatic types with an air-operated vise are available. Special vise insets for holding thin-walled pipe or tubing to prevent distortion are shown in Figure 1-9A. The other holding arrangement is shown in Figure 1-9B. This shows square material being held for optimum cutting angle. Other special shapes can be made to fit in the vises, or they can be purchased from the saw manufacturer.

The semiautomatic machines have heavy-duty feed mechanisms to feed the metal to the saw. The saw can do straight cutting, miter cutting, slot (or longitudinal) cutting, or any number of other arrangements that will fit within the limits of the machine.

Nonferrous Saw

A machine specifically suited for aluminum or nonferrous and hard plastics is shown in Figure 1-10. The machine cuts from the bottom up. The cutting blade speed is 9840 surface feet per minute (sfpm). The workpiece is held from the top by a quick-clamping vise. Maximum saw diameter is 12 inches. A built-in spray mist coolant system provides effective blade lubrication. Mitering is possible in either 45° or 60° and in both directions.

Figure 1-8   Manually operated metal-cutting table saws. *(Courtesy of Kalamazoo Saw from Clausing Industrial Inc.)*

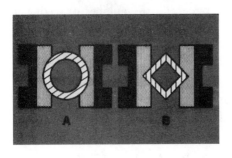

Figure 1-9   Special vise inserts. *(Courtesy of Kalamazoo Saw from Clausing Industrial Inc.)*

Blade

The circular saw blade is the most important part of the saw. It is chosen to do a specific job. The blade is important for the quality of the cut. High-speed blades are coated with a special nitrate treatment to retain strength and hardness, as well as to facilitate coolant penetration. Carbide-tipped blades are made of a special alloy with carbide-tipped teeth for high-speed cutting of aluminum and other light nonferrous materials.

Figure 1-10 High-speed
aluminum and nonferrous saw.

*(Courtesy of Kalamazoo Saw from Clausing
Industrial Inc.)*

Table 1-5 shows the approximate number of teeth for the blade
diameter. This table indicates the actual number of teeth engaged
with the material being cut. No fewer than two to three teeth
should engage the work at all times. No more than seven teeth
should be engaging the work at any one time.

Table 1-6 shows the blade pitch and the number of teeth per
inch. Table 1-7 indicates the blade speed in feet per minute for the
different spindle speeds and blade diameters.

Table 1-8 shows recommended sawing speeds for various materials cut by this type of saw. Table 1-9 indicates the spindle speeds
available for five different models of saws. The spindle speeds for
the circular saws for ferrous metals are 26 to 52 rpm, or a very slow
speed. The nonferrous model saw uses a speed of 3000 rpm, or a
somewhat faster saw blade.

Table 1-5   Approximate Number of TPI

Number of Teeth on Blade	Blade Diameter, in.					
	6	8	10	11	12	14
80	4	3	2½	2	2	1½
100	5	4	3	3	2½	2
120	6	5	4	3½	3	2½

Table 1-5 *(continued)*

Number of Teeth on Blade	Blade Diameter, in.					
	6	8	10	11	12	14
140	7	5½	4½	4	3½	3
160	8	6	5	4½	4	3½
180	9	7	6	5	5	4
200	11	8	6½	6	5½	4½
220	12	9	7	6½	6	5
240	13	10	7½	7	6½	5½
260			8	7½	7	
280			9	8	8	
300			10	9		

Courtesy Startrite, Inc.

Table 1-6 Pitch Selection

Pitch Designation	2	3	4	5	6	8	10	12
Approx. Teeth per Inch	14	10	6	5	4	3	2½	2

Table 1-7 Blade Speeds (fpm)

Spindle Speed, rpm	Blade Diameter, in.					
	8	10	11	12	13	14
26	54	68	75	82	89	95
52	108	136	150	164	177	190
3000				9425		

Table 1-8 Recommended Sawing Guide

Material	Blade Speed, ft/min.
Mild steel	60–165
High-carbon, stainless, alloys, tool steel	40–50
Cast iron	40–125
Bronze	100–170
Brass, copper	120–170
Aluminum	150–9000

Table I-9 Startrite Spindle Speeds

Model	Spindle Speed, rpm
CF 275	26–52
CF 300	26–52
CF 325	26–52
CF 350	23–46
CN 300 T	300

Courtesy Startrite Corp.

Summary

Power hacksaws, power band saws, and circular saws are very important to machine shop operations. A large number of power hacksaws are in use in the metalworking industry.

A power hacksaw is an essential machine in most machine shop operations. It is designed to make the sawing of metal a mechanical operation. High-speed tungsten steel and high-speed molybdenum steel are the most commonly used materials in power saw blades. High-speed, shatterproof blades are designed to meet safety and performance requirements. These blades are designed with 12-inch, 14-inch, and 17-inch lengths with a 1-inch or 1¼-inch width.

In the past few years, the power band saw has become very important in machine shop operations. In some instances it is used in production operations prior to final machining operations. There are three types of saw blades: the regular tooth, skip tooth, and hook tooth. Small and thin-wall sections of metal require a fine-tooth blade. Large metal sections require the use of blades with coarse teeth so that adequate chip clearance is provided. Two teeth should be engaged in the metal to be cut at all times. Soft, easily machined materials require slightly coarser teeth to provide chip clearance. Hard metals of low machinability require finer teeth so that there are more cutting edges per inch. Stainless steel should be sawed with a coarse-tooth saw blade for best results.

Cast iron is always cut dry. Almost all the commercially available soluble oils or light cutting oils will give good results when cutting ferrous metals. When aluminum is cut, paraffin and beeswax are commonly used lubricants. Cutting oil mineral base with fatty oils added is one of four types of cutting fluids. Cutting oil with light viscosity fatty oils is another, as well as synthetic water-soluble cutting agent and soluble oil cutting agent with fatty oils and sulfurized for extreme pressure with antiweldment properties.

The three types of feeds used on power band saws are hand, hydraulic, and mechanical.

Many difficult manufacturing problems have been solved with friction sawing. This process has distinct advantages. Some of the limitations are important. Friction cutting leaves a very heavy burr on the underside of the material being cut, thicknesses over ⅝ inch are extremely difficult to cut, and heat from the friction-sawing process may have a tempering effect on the edges of some materials.

Circular saws have some advantages. They produce a burr-free mill finish when slow-speed cold sawing is used. This eliminates secondary operations on tubing, channels, angles, and solid stock. The rigidity of the blade produces cuts of extreme accuracy and close tolerances.

Review Questions

1. Why are power hacksaws preferred in the machine shop?

2. Brass may be cut at the same speed as ＿＿＿ if a suitable cutting compound is used.

3. List the basic parts of a power hacksaw.

4. What are the three basic types of tooth set on metal-cutting band saw blades?

5. A 1-inch band saw blade can cut with a radius of ＿＿＿＿＿ inches.

6. Why is soluble oil cutting agent water-soluble?

7. Does cast iron need a cutting fluid? Why?

8. What is the probable cause when the blade develops a twist?

9. What is the recommended correction for a twisted blade problem?

10. What are the recommended friction-cutting speeds for gray cast iron?

11. What are three of the limitations to friction cutting?

12. What are the advantages of using circular saws for cutting?

13. How many teeth per inch (TPI) are actually cutting the metal with a 14-inch, 100-tooth blade?

14. Why do some power hacksaws need a coolant system?

15. What type of blades do power hacksaws use?

Chapter 2

Basic Machine Tool Operations

Machine tools are relatively new. The machine tools in use today are the products of the Industrial Revolution, which started a little more than 200 years ago. Prior to the Industrial Revolution, the various items that people wore or used were made by hand tools that for centuries had changed little in their basic design. It was not until machine tools replaced hand tools that the life of abundance we have come to know and enjoy began to evolve.

Machine power was substituted for muscle power to cut, shape, and form metal products. Drilling, turning, milling, planing, grinding, and metal forming, the basic machine tool processes in use today, evolved from hand tools.

Prior to World War II, machine tools played a very important part in making the United States the most highly industrialized nation on the face of the earth. Along with this came one of the highest standards of living in the world. During World War II, the country's machine tool population more than doubled to meet the war production demands.

In the 1950s, the design of machine tools began to change drastically. These changes were primarily the result of the development of numerically controlled machine tools and the advancements made in the area of electronics. In the short time since then, a technological revolution has taken place in the design and development of machine tools. Advances in design and development have been so great that they have surpassed all previous advancements.

The development of this new technology did not occur by accident. This new technology is the direct result of an unprecedented upsurge in spending by the machine tool industry for research and development. The rate of technological change is accelerating more and more rapidly every year. The life span of a new machine tool is considerably shorter than it once was because the speed with which new products are constantly being developed has rapidly increased.

Once machine tools were simple and uncomplicated. This is no longer true. Machine tools are becoming more and more complex as more features are being built into their design and performance. For the machine tools we hear so much about, see Figures 2-1 through 2-8.

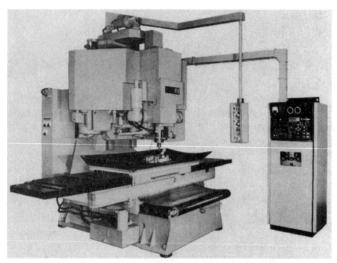

Figure 2-1 A vertical spindle machining center with heavy milling capacity. This milling capacity, together with accurate boring, drilling, and tapping, expands the capability of this type of machine to accommodate an exceptionally broad range of parts. Once set up, the operation is completely automatic, with fast recycling, no lost motion, and no unnecessary idle time. *(Courtesy Monarch Cortland.)*

Figure 2-2 The automatic tool changer on the vertical spindle machining center. Tools are stored in pockets connected to an endless chain conveyor. *(Courtesy Monarch Cortland.)*

Figure 2-3   Tool in cut, quill down, stored tools in covered enclosure.
(Courtesy Monarch Cortland.)

Figure 2-4   Quill up; fingers grasp the adapter; cover opens, exposing the tool storage conveyor. *(Courtesy Monarch Cortland.)*

Figure 2-5 Tool leaves the spindle. *(Courtesy Monarch Cortland.)*

Figure 2-6 The tool swings in the arc. *(Courtesy Monarch Cortland.)*

Figure 2-7   The tool is in storage. *(Courtesy Monarch Cortland.)*

Figure 2-8   The next tool starts to spindle. *(Courtesy Monarch Cortland.)*

Today, there are unlimited varieties and combinations of machine tools. Some of these machine tools are small enough to be mounted on a table. Some machine tools are so large that they require special buildings to house them. Their cost can range from a few hundred dollars to hundreds of thousands of dollars. Some machine tools, such as presses, weigh several hundred tons. When some of the machine tools are installed, their foundations have to be extended into the ground to a depth of more than 100 feet. They can also be several stories in height. Some machine tools are incorporated into production lines hundreds of feet in length.

Whether machine tools are small or large, inexpensive or extremely expensive, they can be grouped into six major classifications. These major classifications are based upon the operations performed by the machines to shape metal. These basic operations include the following:

- Drilling and boring (including reaming and tapping, turning, and milling)
- Planing (including shaping and broaching)
- Grinding (including honing and lapping)
- Metal forming (including shearing, stamping, pressing, and forging)

In addition to the basic metal-shaping operations, newer metal-shaping operations have been developed during the past two decades that employ the various characteristics of chemicals, electricity, magnetism, liquids, explosives, light, and sound.

Whether a machine tool is simple or complex, it performs one or more of these operations. Variations of the six basic operations are employed to meet special situations. For example, there are machine tools that combine two or more operations, as in a boring, drilling, and milling machine; a stamping, punching, and shearing press; or a combination milling and planing machine.

Drilling and Boring

Drilling is a basic machine shop operation dating back to primitive humans. It consists of cutting a round hole by means of a rotating drill (Figure 2-9). *Boring*, on the other hand, involves the finishing of a hole already drilled or cored. This is accomplished by means of a rotating, offset, single-point tool that somewhat resembles the tool used in a lathe or a planer. The tool is stationary and the work revolves on some boring machines. On other types of boring machines, the reverse is true.

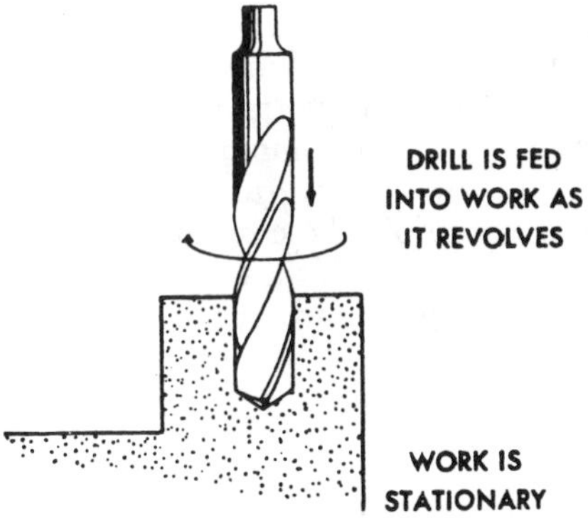

Figure 2-9   Drilling.

DRILL IS FED
INTO WORK AS
IT REVOLVES

WORK IS
STATIONARY

Two other types of machine tools are included under the classification of drilling and boring. These machine tools perform reaming and tapping operations. *Reaming* consists of finishing an already drilled hole. This is done to very close tolerances. *Tapping* is the process of cutting a thread inside a hole so that a screw may be used in it.

Turning

Turning is done on a lathe. The lathe, as the turning machine is commonly called, is the father of all machine tools. The principle of turning has been known since the dawn of civilization, probably originating as the potter's wheel. In the turning operation, the piece of metal to be machined is rotated and the cutting tool is advanced against it (Figure 2-10).

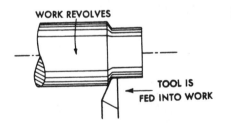

WORK REVOLVES

Figure 2-10   Turning.

TOOL IS
FED INTO WORK

By contrast, the *turret lathe* is a lathe equipped with a multisided toolholder called a *turret*, to which a number of different cutting tools are attached. This device makes it possible to bring several different cutting tools into successive use and to repeat the sequence of machining operations without the need to reset the tools. The cutting tools themselves are mounted and protrude from the turrets.

Single- and multiple-spindle automatics are used when the number of identical parts to be turned is increased from a few to hundreds, or even thousands. These machines perform as many as six or eight different operations at one time on a number of different parts. Single- and multiple-spindle automatics are entirely automatic. Once set up and put into operation, these machines relieve the operator of all but two duties. The duties of the operator consist of monitoring operations and gauging the accuracy of finished parts.

Milling

Milling consists of machining a piece of metal by bringing it into contact with a rotating cutting tool with multiple cutting edges (Figure 2-11). A narrow milling cutter resembles a circular saw blade familiar to most people. Other milling cutters may have spiral edges, which give the cutter the appearance of a huge screw.

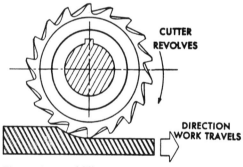

CUTTER
REVOLVES

DIRECTION
WORK TRAVELS

Figure 2-11   Milling.

There are many other milling machines designed for various kinds of work. For example, the *planer type* is built like a planer, but it has multiple-tooth revolving cutters. Machines that use the milling principle but are built especially to make gears are called *hobbing machines*. Some of the shapes produced on milling machines are extremely simple (like the slots and flat surfaces produced by circular saws). Other shapes are more complex and may consist of a variety of combinations of flat and curved surfaces, depending upon the shape of the cutting edges of the tool and the path of travel of the tool.

Planing

Planing or shaping metal with a machine tool is a process somewhat similar to planing wood with a carpenter's hand plane. The essential difference lies in the large size of the machine tool and the fact that it is not portable. The cutting tool remains in a fixed position, while the work is moved back and forth beneath it.

On a shaper the process is reversed. The workpiece is held stationary, while the cutting tool travels back and forth (Figure 2-12).

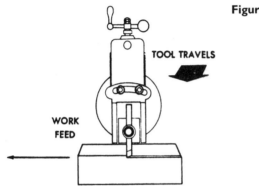

Figure 2-12 Shaping.

TOOL TRAVELS

WORK
FEED

A somewhat similar operation is known as slotting. This operation is performed vertically. Slotters, or vertical shapers, are used principally to cut certain types of gears.

Broaches may be classified as planing machines. The broach has a number of cutting teeth. Each cutting edge is a little higher than the previous one, and it is graduated to the final size required. The broach is pulled or pushed over the surface to be finished. It may be applied internally (for example, to finish a square hole) or externally (for example, to produce a flat surface or a special shape).

Planers are usually very large. Sometimes they are large enough to handle the machining of surfaces that are 15 to 20 feet wide and about twice as long.

Grinding

Grinding consists of shaping a piece of work by bringing it into contact with a revolving abrasive wheel (Figure 2-13). The process is often used for the final finishing to close dimensions of a part that has been heat-treated to make it very hard. In recent years, grinding has found increasing applications in heavy-duty metal removal,

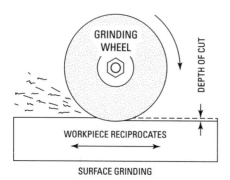

Figure 2-13 Grinding.

replacing machines with cutting tools. This process is referred to as *abrasive machining.*

The grinding machine can correct distortions that have resulted from the heat treatment process. It may be used on external cylindrical surfaces, in holes, for flat surfaces, and for threads.

Under the classification of grinding are included operations known as *lapping* and *honing.*

- Lapping involves the use of abrasive pastes and compounds. It is limited in its use to extremely small amounts of stock removal and to situations where there is a high degree of precision and surface finish needed.

- The honing technique, in contrast, is widely accepted as a process separate from that of lapping. For example, there are honing machines with rotating heads that carry abrasive inserts for the extremely accurate finishing of holes.

Metal forming includes shearing, stamping, pressing, and forging metals of many kinds. It requires the use of many kinds of tools, including the following:

1. *Shear*—This tool is used to cut metal into the required shapes.

2. *Punch press*—This tool is used to punch holes in metal sheet and plate.

3. *Mechanical press*—This tool is used to blank out the desired shape from a metal sheet and squeeze it into the final shape in a die under tremendous pressure.

4. *Hydraulic press*—This tool does the same work as the mechanical press by the application of hydraulic power.

5. *Drop hammer*—This tool is operated by steam or air. It is used to forge or hammer white-hot metal on an anvil.

6. *Forging machine*—This tool squeezes a piece of white-hot metal under great pressure in a die. During the process, the metal flows into every part of the die cavity where it assumes the shape of the cavity.

The research and development efforts of the past few years have resulted in a number of new operations for shaping metal into useful parts. Discounting entirely those processes that form metal in its molten state or in a powder state, many new developments have the effect of broadening the capabilities of machine tools.

Following are some of the better-known developments:

- Abrasive machining
- Capacitor discharge machining
- Cold extrusion
- Combustion machining
- Electrical discharge machining
- Electromechanical machining
- Electrolytic machining
- Electrospark forming
- Electron beam machining
- Explosive forming
- Fission fragment exposure and etching
- Gas forming
- Hot machining
- Hydroforming
- Laser (light beam) cutting
- Magnetic forming
- Plasma machining
- Spark forming
- Ultrasonic cutting and forming

A number of the processes were developed to do specific work, such as machining extremely hard materials. Many of the developments came about through aircraft, atomic energy, and rocket research. Some of these processes have been adapted to production-type machines, as in electrochemical milling, electrical discharge machining, magnetic forming, abrasive machining, electron beam machining, electrospark forming, and cold extrusion. A number of processes are still awaiting the basic research necessary to incorporate a process into a piece of production equipment. Although some of these processes are today's curiosities, they will play an important part in manufacturing goods for a rapidly growing economy.

Summary

Machines tools are relatively new. The machine tools in use today are the products of the Industrial Revolution, which started a little more than 200 years ago.

Machine power was substituted for muscle power to cut, shape, and form metal products. Drilling, turning, milling, planing, grinding, and metal forming, the basic machine tool processes in use today, evolved from hand tools.

Drilling is the basic machine shop operation and dates back to primitive humans. It consists of cutting a round hole by means of a rotating drill.

Turning is done on a lathe. The lathe, as the turning machine is commonly called, is the father of all machine tools. In the turning operation, the piece of metal to be machined is rotated and the cutting tool is advanced against it.

Milling consists of machining a piece of metal by bringing it into contact with a rotating cutting tool with multiple cutting edges. There are many different types of milling machines.

Planing or shaping metal with a machine tool is a process somewhat similar to planing wood. The cutting tool remains in a fixed position while the work is moved back and forth beneath it.

Grinding consists of shaping a piece of work by bringing it into contact with a revolving abrasive wheel. Grinding can correct distortions that have resulted from the heat treatment process. It may be used on external cylindrical surfaces, in holes, for flat surfaces, and for threads.

Metal forming includes shearing, stamping, pressing, and forging metals of many kinds. It requires the use of many kinds of tools. Basically, they are the shear, the punch press, the mechanical press, the hydraulic press, the drop hammer, and the forging machine.

The research and development efforts of the past few years have resulted in a number of new operations for shaping metal into useful parts. These new operations will be used more and more in the years ahead. Recent developments in robotics and electronic computers and their application to this field have produced some rather sophisticated machines capable of unbelievable tasks.

Review Questions

1. What is drilling?
2. What is turning?
3. What is milling?
4. What is planing?
5. What is grinding?
6. What is metal forming?
7. What is meant by the machine tool industry?
8. What is a spindle?
9. What are the classifications of machine tools as grouped by the operations they perform?
10. How can turning be done on a lathe?
11. How is a turret lathe different from a machine lathe?
12. How does the planer work?
13. What is a shaper?
14. How are broaches classified?
15. Describe lapping and honing.

Chapter 3

Drilling Machines

Round holes are commonly drilled in metal by means of a machine tool called a *drill press*. The term *drilling machines* is much broader in meaning and includes all types of machines designed for drilling holes into metal.

Many operations other than drilling a round hole can be performed on the drill press. Some of these are sanding, counterboring and countersinking, honing, reaming, lapping, and tapping. Considerable skill is required to drill a hole of proper size in exactly the desired location at a high rate of production. The machine operator must be able to locate the hole properly and accurately, and the machine operator must be able to align the drill correctly.

Basic Construction

Successful operation of the drill press requires the operator to be familiar with all parts and to have an adequate working knowledge of the machine itself. The operator must also be able to set up the work properly, to select proper speeds and feeds, and to use the correct coolant.

The bench-type drill press (Figure 3-1) and the floor-type drill press (Figure 3-2) are commonly found in home workshops and in industrial machine shops. These machines are designed to rotate a cutting tool (twist drill, countersink, counterbore, and so on) to advance the cutting tool into the metal and to support the workpiece.

Head

The design of the drill press *head* varies with different machines (Figure 3-3). In most machines, the electric motor is bolted to the head, and a V-belt drive is used to drive the spindle at from three to five different speeds by shifting the V-belt from step-to-step on the pulleys. For maximum life of the V-belt, the belt tension should be just tight enough that loosening is not necessary to shift it.

Spindle

The *spindle* is the rotating part. It is usually splined and made of alloy steel. The spindle rotates and moves up and down in a quill or sleeve, which slides on bearings. A pinion engages a rack fastened to the quill to provide vertical movement of the quill, permitting the

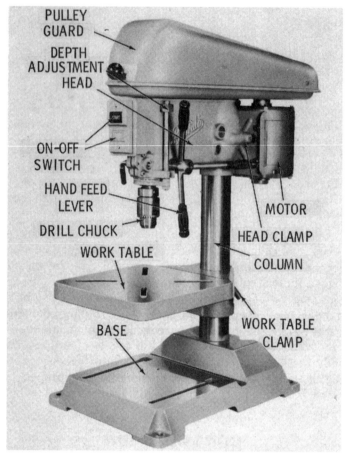

PULLEY
GUARD
DEPTH
ADJUSTMENT
HEAD

ON-OFF
SWITCH

HAND FEED
LEVER

DRILL CHUCK

WORK TABLE

MOTOR
HEAD CLAMP
COLUMN

BASE

WORK TABLE
CLAMP

Figure 3-1   Bench-type sensitive drill press. *(Courtesy Buffalo Forge.)*

twist drill to be either fed into or withdrawn from the workpiece. A typical spindle assembly is shown in Figure 3-4.

Spindle speed is controlled on smaller machines by changing the V-belt from one step to another on the pulleys. Gearboxes are provided on the larger machines for making changes in spindle speed.

Some upright drill presses have *back gears*, which supply more power to the spindle. Slower spindle speeds are a result of this increase in power. On some machines, the back gears can be

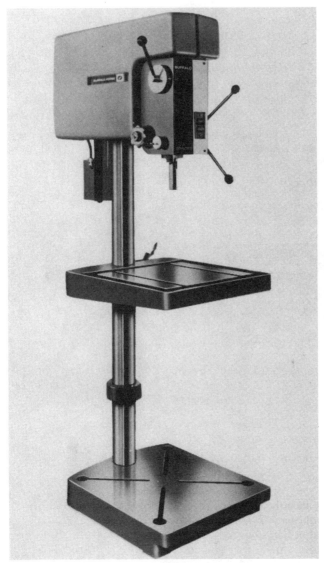

Figure 3-2 Floor-type sensitive drill press. *(Courtesy Buffalo Forge.)*

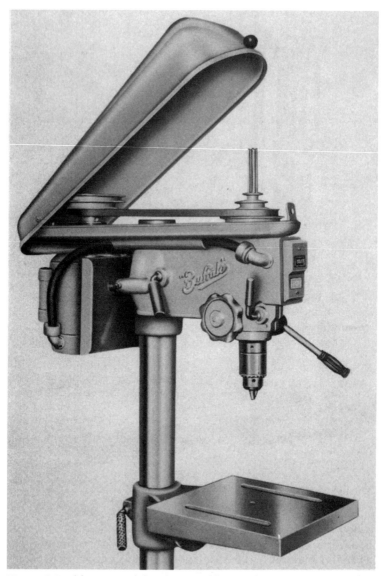

Figure 3-3   Newer models use a variable speed arrangement that changes pulley sizes from a control in front of the drill press.

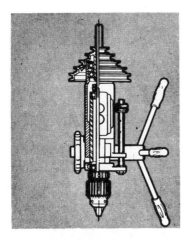

Figure 3-4 A typical spindle assembly for a drill press.
(Courtesy Buffalo Forge.)

engaged and disengaged by means of a lever conveniently located on the machine.

Table

The *table* is supported on the column of the drill press. It can be moved both vertically and horizontally to the desired working position, or it can be swung around so that it will be out of the way. Most tables are slotted so that the work, or a drill press vise for holding the work, can be bolted to them (Figure 3-5).

Figure 3-5 A drill press vise used for holding the work on the drill press table. *(Courtesy Ridge Tool Co.)*

Base

The supporting member of the entire drill press structure is the *base*. It is a heavy casting with holes or slots for bolting it to the bench and for securing the work or workpiece directly to the base. The base supports the column, which in turn supports the table and head.

Feed

The *feed* on a drill press can be either manual or automatic. Feed is expressed in thousandths of an inch per revolution of the spindle. The feed of a twist drill is the distance the drill moves into the work per revolution of the spindle. Automatic feed is always referred to as *power feed*. A hand lever or handwheel is used on manual-type drill presses in which small twist drills are mounted for light work.

Capacity

Capacity of a drill press is expressed in several ways. Usually, drill presses are rated by the distance from the center of the twist drill to the column of the machine (Figure 3-6). Some drill presses are rated by the distance from the top of the table in its lowest position to the tip of the spindle at its highest position. Another rating is given as the distance the spindle can travel from its uppermost position to its lowest position. Still another rating is the largest straight shank

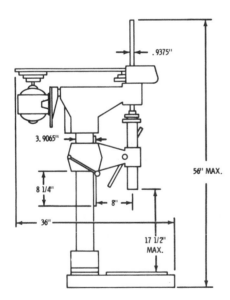

Figure 3-6 Note the capacity of the bench-type sensitive drill.

twist drill that can be mounted in the drill press spindle or drill chuck.

Types of Drilling Machines

Several types of drilling machines are in existence. *Manual-feed drill presses* are either light-duty or medium-duty machines. Those with automatic or power feed are heavy-duty machines. Drilling machines are sometimes classified as either *vertical-spindle* or *horizontal-spindle machines* and as either *single-spindle* or *multi-spindle machines*. The multispindle machines are also called gang drilling machines.

Sensitive Drill Press

These machines can be of either the bench type (see Figure 3-1) or the floor type (see Figure 3-2). They are belt-driven, hand-fed drill presses. The hand feed of the sensitive drill press permits the operator to "feel" the cutting action at the end of the twist drill. A counterbalanced spindle is moved vertically with a hand lever or handwheel. These machines are designed for relatively light work with small twist drills, which are prone to breakage under power feed.

Twist drills up to ½ inch in diameter can be used on the sensitive drill press. The end of the spindle is bored for a standard No. 2 Morse taper, to fit the tapered shank of a drill chuck or a twist drill.

Variable-Speed Upright Drill Press

The *back-geared drill press* (Figure 3-7 and Figure 3-8) has a greater range of speeds than the floor-type sensitive drill press. It is also larger and equipped with a power feed. The reversing mechanism on the back-geared upright drill press also permits tapping operations.

Radial Drill Press

In a *radial drill press* (Figure 3-9), the vertical spindle can be positioned horizontally and locked on an arm that can be swiveled about, and raised and lowered on a vertical column. Thus, the spindle can be placed in any position within its range. The various arm and spindle movements are shown in Figure 3-10. Some radial drills do not have the last two movements illustrated.

Because the drilling head is moveable to any position, it is not necessary to move heavy work for each hole that is to be drilled. Therefore, the radial drill is especially adaptable for work of this kind. The radial drill is a heavy-duty drilling machine. It is capable

Figure 3-7 A view of a drill press head with the guard in the open position. *(Courtesy Buffalo Forge.)*

Figure 3-8 Variable-speed drill press. Speed is adjusted by control handle on top at eye level.

of handling work that, because of its weight or size, cannot be mounted on a drill press table.

A radial drill can be equipped with either a plain table or a universal tilting table (Figure 3-11). The universal tilting table is designed for angular drilling. The table can be rotated through 360°. When several drills are mounted on the same table, the machine is called a gang drilling machine (Figure 3-12).

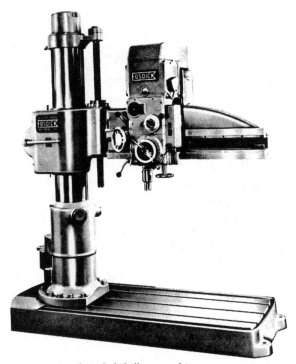

Figure 3-9 A radial drilling machine. *(Courtesy Buffalo Forge.)*

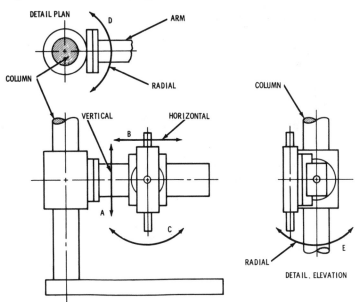

Figure 3-10 Note the movements of a radial drill having unlimited application or motion.

Figure 3-11 Radial drill press tables. The plain table is shown (left). The universal or tilting table (right) can be rotated through 360° and tilted through 90°. *(Courtesy Giddings and Lewis Machine Tool Co.)*

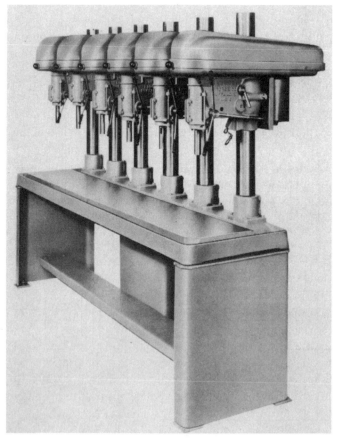

Figure 3-12 A bench-type, six-spindle gang drilling machine.
(Courtesy Buffalo Forge.)

Summary

A drill press is a machine commonly used to drill round holes in metal or wood. Many operations other than drilling round holes can be performed on a drill press. Some of these are sanding, counterboring, countersinking, honing, reaming, and tapping.

Various parts of the drill press are the head, spindle, table, base, and feed. The head generally houses the drill motor and the V-belt drive assembly. The spindle is the rotating part of the drill and slides up and down in a sleeve, which slides on a bearing. The spindle also holds the drill chuck, which permits the twist drill to be either fed or withdrawn from the workpiece. The table is supported on a column of the drill press. The table can be moved both vertically and horizontally to the desired working position, or it can be swung around so that it will be out of the way. The table also supports the work and is generally slotted so the work can be clamped or bolted down to prevent movement.

The feed can be either manual or automatic. Automatic feed is always referred to as power feed. The feed of a drill is the distance the drill moves into the work.

Successful operation of the drill press requires the operator to be familiar with all parts. The operator must have an adequate working knowledge of the machine itself. The operator must also be able to set up the work properly, to select proper speeds and feeds, and to use the correct coolant. There are two types of drill presses, the bench type and the floor type.

The radial drill press can be equipped with either a plain table or a universal tilting table. The universal tilting table is designed for angular drilling. The table can be rotated through 360°. When several drills are mounted on the same table, the machine is called a gang drilling machine.

Review Questions

1. Explain the purpose of the drill press.
2. Name the basic parts of a drill press.
3. What is meant by the capacity of a drill press?
4. What is a radial drill press?
5. How is spindle speed controlled in a drill press?
6. Why is spindle speed important in a drill press?
7. Why do you need a drill press vise?
8. What is the advantage of a back-geared motor drive on a drill press?
9. What is the advantage of a six-spindle drill press?
10. Where are six-spindle drill presses used?

Chapter 4

Drilling Machine Operations

Many operations can be performed on the drilling machine. The setup for any drilling machine operation should be carefully studied and checked before proceeding with the operation.

Drilling Machine Operations

Good sound judgment and trial-and-error are important in each drill press operation. Hard and fast rules are not practical for operation of the drill press because composition and hardness of material, type of machine, condition of the machine, condition of the cutting tool, lubricant, depth of hole, and many other factors influence the speed and feed at which a material can be worked. However, suggestions can be used as a guide, and the operator can make intelligent observations and adjustment of feeds and speeds for a given operation.

Drilling

The chief operation performed on the drill press is *drilling*, which is the removal of solid metal to form a circular hole. Prior to drilling a hole in metal, the hole is located by drawing two lines at right angles, and a center punch is used to make an indentation for the drill point at the center to aid the drill in getting started (Figure 4-1).

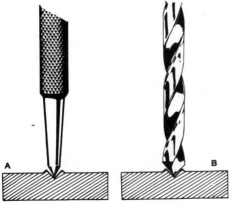

Figure 4-1 Using the center punch. (A) Center punch used to make an indentation in the workpiece. (B) Indentation should be large enough for the point of the drill.

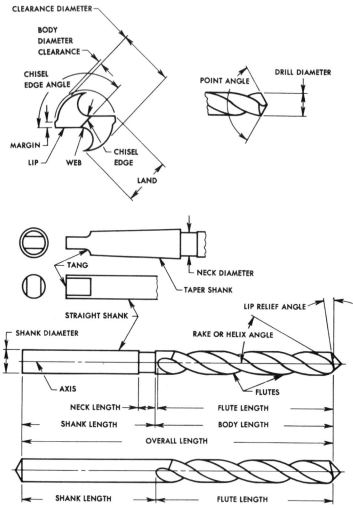

Figure 4-2 Elements of twist drills. *(Courtesy Aluminum Co. of America.)*

Never attempt to start a twist drill without first using a center punch to make the indentation for starting the drill point.

Twist Drill
Twist drills are used in drilling machine operations. They are usually made from carbon steel or high-speed steel. Carbon steel twist drills are not suited for high-speed production work. They must be

operated at lower cutting speeds than twist drills made from high-speed steel.

All twist drills have three major parts: the shank, the body, and the point. They are available in many different sizes (see Table 4-1). These different sizes are classified as follows:

1. Fractional: $\frac{1}{64}''$ to $4'' \times \frac{1}{64}''$ increments
2. Millimeter
3. Number: 1 to 60, 61 to 80
4. Letter: A to Z

Table 4-1   Decimal and Metric Equivalents of Number, Letter, and Fractional Size Drills

Drill Size	Decimal	mm	Drill Size	Decimal	mm
80	0.0135	0.3429	47	0.0785	1.9939
79	0.0145	0.3683	46	0.0810	2.0574
$\frac{1}{64}$	0.0156	0.3969	45	0.0820	2.0828
78	0.0160	0.4064	44	0.0860	2.1844
77	0.0180	0.4572	43	0.0890	2.2606
76	0.0200	0.5040	42	0.0935	2.3749
75	0.0210	0.5334	$\frac{3}{32}$	0.0938	2.3812
74	0.0224	0.5715	41	0.0960	2.4384
73	0.0240	0.6096	40	0.0980	2.4892
72	0.0250	0.6350	39	0.0995	2.5273
71	0.0260	0.6604	38	0.1015	2.5781
70	0.0280	0.7072	37	0.1040	2.6416
69	0.0292	0.7417	36	0.1065	2.7051
68	0.0310	0.7874	$\frac{7}{64}$	0.1094	2.7781
$\frac{1}{32}$	0.0313	0.7937	35	0.1100	2.7940
67	0.0320	0.8128	34	0.1110	2.8194
66	0.0330	0.8382	33	0.1130	2.8702
65	0.0350	0.8890	32	0.1160	2.9464
64	0.0360	0.9144	31	0.1200	3.0480
63	0.0370	0.9398	$\frac{1}{8}$	0.1250	3.1750
62	0.0380	0.9652	30	0.1285	3.2659
61	0.0390	0.9906	29	0.1360	3.4544
60	0.0400	1.0160	28	0.1405	3.5687
59	0.0410	1.0414	$\frac{9}{64}$	0.1406	3.5719

(continued)

Table 4-1 (continued)

Drill Size	Decimal	mm	Drill Size	Decimal	mm
58	0.0420	1.0668	27	0.1440	3.6596
57	0.0430	1.0922	26	0.1470	3.7338
56	0.0465	1.1811	25	0.1495	3.7773
$3/64$	0.0469	1.1906	24	0.1520	3.8608
55	0.0520	1.3208	23	0.1540	3.9116
54	0.0550	1.3970	$5/32$	0.1562	3.9687
53	0.0595	1.5113	22	0.1570	3.9878
$1/16$	0.0625	1.5825	21	0.1590	4.0386
52	0.0635	1.6129	20	0.1610	4.0894
51	0.0670	1.7018	19	0.1660	4.2164
50	0.0700	1.7780	18	0.1695	4.3053
49	0.0730	1.8542	$11/64$	0.1719	4.3656
48	0.0760	1.9304	17	0.1730	4.3942
$5/64$	0.0781	1.9844	16	0.1770	4.4958
15	0.1800	4.5720	$21/64$	0.3281	8.3344
14	0.1820	4.6228	Q	0.3320	8.4328
13	0.1850	4.6990	R	0.3390	8.6106
$3/16$	0.1875	4.7625	$11/32$	0.3438	8.7312
12	0.1890	4.8006	S	0.3480	8.8392
11	0.1910	4.8514	T	0.3580	9.0932
10	0.1935	4.8149	$23/64$	0.3594	9.1281
9	0.1960	4.9784	U	0.3680	9.3472
8	0.1990	5.0546	$3/8$	0.3750	9.5250
7	0.2010	5.1054	V	0.3770	9.5758
$13/64$	0.2031	5.1594	W	0.3860	0.8044
6	0.2040	5.1816	$25/64$	0.3906	9.9219
5	0.2055	5.2197	X	0.3970	10.0380
4	0.2090	5.3086	Y	0.4040	10.2616
3	0.2130	5.4102	$13/32$	0.4062	10.3187
$7/32$	0.2188	5.5562	Z	0.4130	10.4902
2	0.2210	5.6134	$27/64$	0.4219	10.7156
1	0.2280	5.8012	$7/16$	0.4375	11.1125
A	0.2340	5.9436	$29/64$	0.4531	11.5094
$15/64$	0.2344	5.9531	$15/32$	0.4688	11.9062
B	0.2380	6.0452	$31/64$	0.4844	12.3031
C	0.2420	6.1468	$1/2$	0.5000	12.7000

Table 4-1 *(continued)*

Drill Size	Decimal	mm	Drill Size	Decimal	mm
D	0.2460	6.2484	$^{33}/_{64}$	0.5156	13.0969
E¼	0.2500	6.3500	$^{17}/_{32}$	0.5313	13.4937
F	0.2570	6.5278	$^{35}/_{64}$	0.5489	13.8906
G	0.2610	6.6294	$^{9}/_{16}$	0.5625	14.2875
$^{17}/_{64}$	0.2656	6.7469	$^{37}/_{64}$	0.5781	14.6844
H	0.2660	6.7564	$^{19}/_{32}$	0.5938	15.0812
I	0.2720	6.9088	$^{39}/_{64}$	0.6094	15.4781
J	0.2770	7.0358	$^{5}/_{8}$	0.6250	15.8750
K	0.2810	7.1374	$^{41}/_{64}$	0.6406	16.2719
$^{9}/_{32}$	0.2812	7.1437	$^{21}/_{32}$	0.6562	16.6687
L	0.2900	7.3660	$^{43}/_{64}$	0.6719	17.0656
M	0.2950	7.4930	$^{11}/_{16}$	0.6875	17.4625
$^{19}/_{64}$	0.2969	7.5406	$^{45}/_{64}$	0.7031	17.8594
N	0.3020	7.6508	$^{23}/_{32}$	0.7188	18.2562
$^{5}/_{16}$	0.3125	7.9375	$^{47}/_{64}$	0.7344	18.6531
O	0.3160	8.0264	$^{3}/_{4}$	0.7500	19.0500
P	0.3230	8.2042	$^{49}/_{64}$	0.7656	19.4469
$^{25}/_{32}$	0.7812	19.8437	$^{29}/_{32}$	0.9062	23.0187
$^{51}/_{64}$	0.7969	20.2406	$^{59}/_{64}$	0.9219	23.4156
$^{13}/_{16}$	0.8125	20.6375	$^{15}/_{16}$	0.9375	23.8125
$^{53}/_{64}$	0.8281	21.0344	$^{61}/_{64}$	0.9531	24.2094
$^{27}/_{32}$	0.8438	21.4312	$^{32}/_{32}$	0.9688	24.6062
$^{55}/_{64}$	0.8594	21.8281	$^{63}/_{64}$	0.9844	25.0031
$^{7}/_{8}$	0.8750	22.2250	1	1.0000	25.4001
$^{57}/_{64}$	0.8906	22.6219			

There are many different kinds of twist drills. Some twist drills are designed for specific types of work. A number of different twist drills are illustrated, with their specific applications, in Figures 4-3 through 4-13.

Figure 4-3 Taper-length, tanged, automotive series, straight-shank twist drill regularly furnished with tangs for use with split-sleeve drill drivers. *(Courtesy National Drill and Tool Co.)*

Figure 4-4 Straight-shank, three-flute core drill. *(Courtesy National Drill and Tool Co.)*

Figure 4-5 Straight-shank, four-flute core drill. *(Courtesy National Drill and Tool Co.)*

Figure 4-6 Straight-flute drill for free-machining brass, bronze, or other soft materials, particularly on screw machines. Also suitable for drilling thin sheet material, because of lack of tendency to "hog." *(Courtesy Morse Twist Drill and Machine Co.)*

Figure 4-7 Drill used principally for drilling molded plastics. Also used for drilling hard rubber, wood, and aluminum and magnesium alloys. *(Courtesy National Drill and Tool Co.)*

Figure 4-8 Low-helix drill used extensively in screw machines making parts from screw stock and brass. *(Courtesy National Drill and Tool Co.)*

Figure 4-9 High-helix drill designed with higher cutting rake and improved chip conveying properties for use on materials such as aluminum, die-casting alloys, and some plastics. *(Courtesy National Drill and Tool Co.)*

Figure 4-10 Screw-machine-length straight-shank twist drill. *(Courtesy National Drill and Tool Co.)*

Figure 4-11   Center drill. *(Courtesy National Drill and Tool Co.)*

Figure 4-12   Starting drill. *(Courtesy National Drill and Tool Co.)*

Figure 4-13   Oil-hole twist drills for production work in all types of materials on screw machine or turret lathes. *(Courtesy National Drill and Tool Co.)*

Before any twist drill is used, care must be taken to see that the twist drill is properly ground. Figures 4-14 through 4-19 illustrate the correct and incorrect procedures.

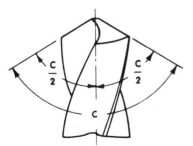

Figure 4-14   The two cutting edges (lips) should be equal in length and should form equal angles with the axis of the drill. Angle C should be 135° for drilling hard or alloy steels. For drilling soft materials and for general purposes, angle C should be 118°.
(Courtesy National Drill and Tool Co.)

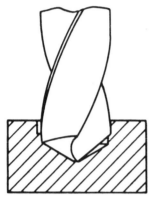

Figure 4-15 A twist drill with the lips ground at unequal angles with the axis of the drill can be the cause of an oversized hole. Unequal angles also result in unnecessary breakage and cause the drill to dull quickly. *(Courtesy National Drill and Tool Co.)*

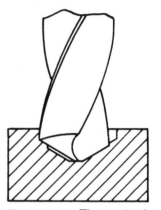

Figure 4-16 The result of grinding the drill with equal angles but with the lips unequal in length. *(Courtesy National Drill and Tool Co.)*

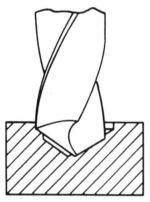

Figure 4-17 The result of grinding the drill with lips of unequal angles and unequal lengths. *(Courtesy National Drill and Tool Co.)*

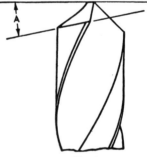

Figure 4-18 The lip relief angle A should vary according to the material to be drilled, and according to the diameter of the drill. Lesser relief angles are required for hard and tough materials than for soft, free-machining materials. *(Courtesy National Drill and Tool Co.)*

Drilling Suggestions

The operator should check to make certain that the drill point has started properly before the drill has cut too deeply. Frequently, the drill point fails to seat properly in the punch mark, and a small hole is made off center.

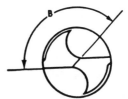

Figure 4-19   The chisel-point angle B increases or decreases with the relief angle and should range from 115° to 135°. *(Courtesy National Drill and Tool Co.)*

The operator should use layout lines and layout circles so that he can determine whether the twist drill is making a hole concentric with the layout circle at the start and is going properly. If the small hole made by the drill is not concentric with the layout circle, the drill should be withdrawn. A cape chisel can be used to cut a groove, or several grooves, on the side toward which the drill should be moved so that the drill can be "drawn" back to the center of the layout circle. It is too late to shift the twist drill point without marring the workpiece, after the drill has begun to cut its full diameter.

After the mounted work is placed on the worktable of the drill press, the worktable should be adjusted to a convenient height for drilling, considering both the length of the twist drill and drill-holder. The height of the spindle should be adjusted to provide the shortest distance for feed of the twist drill to the work.

After the proper speed has been determined, the machine can be started and the twist drill lowered to the work. The drill should be fed to the work slowly until the metal has been penetrated by the point of the drill. Then, a check should be made to see that the drill has started properly.

Feed and Speed

In drilling holes with a small diameter, the danger of twist drill breakage is very great unless the feed and speed are given careful attention, especially at the moment the point of the drill breaks through the other side of the work (Figure 4-20). In general, high speed and light feed are recommended. It is better to err on the side of too much speed than to err on the side of too much feed (except for cast iron, which permits an unusually heavy feed). Speed can be increased to the point where the outside corners of the twist drill show signs of wearing away. Speed can then be reduced and maintained at the reduced speed. Recommended drilling speeds can be obtained from Table 4-2.

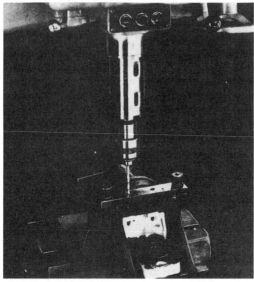

Figure 4-20 Performing small-hole drilling on the Bickford radial drilling machine. Drills and taps as small as ³⁄₁₆ inch (0.1875) and ¼ inch (0.250) can be handled at high speeds. *(Courtesy Giddings and Lewis Machine Tool Co.)*

Table 4-2 **Recommended Drilling Speeds for Materials (High-Speed Drills)**

Material	Recommended Speed in Surface Feet per Minute, ft^2/min.
Aluminum and alloys, brass and bronze, soft	200–300
Bakelite	100–150
Plastics	100–150
Bronze, high tensile	70–100
Cast iron, chilled, steel, stainless (hard)	30–40
Cast iron, hard	70–100
Cast iron, soft	100–150
Magnesium and alloys	250–400
Malleable iron	80–90
Monel, metal	40–50
Nickel	40–60

Table 4-2 (continued)

Material	Recommended Speed in Surface Feet per Minute, ft²/min.
Steel, annealed (.4 to .5 percent C)	60–70
Steel, forgings, wrought iron, steel, tool	50–60
Steel, machine (.2 to .3 percent C)	80–110
Steel, manganese (15 percent Mn), slate, marble	15–25
Steel, soft	80–100
Steel, stainless (free machining)	60–70
Wood	300–400

Too much speed is indicated if the twist drill chips out at the cutting edge (Table 4-3). This is a certain indication of either too heavy feed or too much lip clearance (Figure 4-21). If the twist drill splits up the web, too much feed for the amount of lip clearance is indicated. Either the feed should be decreased or the lip clearance should be increased—or both actions might be used to remedy the difficulty. Also, too much lip clearance at the center (or at any other

Figure 4-21 Using a Bickford radial drilling machine to drill a 1³³⁄₆₄-inch (1.5156) hole in a welded all-steel frame for a 3-ton electric hoist. Feed is 0.025 inch per revolution at 175 rpm. *(Courtesy Giddings and Lewis Machine Tool Co.)*

Table 4-3   Cutting Speeds (Fractional Drills)

Diameter, in.													

Feet per Min.

Diameter, in.	30	40	50	60	70	80	90	100	110	120	130	140	150
	Revolutions per Minute (rpm)												
1/16	1833	2445	3056	3667	4278	4889	5500	6111	6722	7334	7945	8556	9167
1/8	917	1222	1528	1833	2139	2445	2750	3056	3361	3667	3973	4278	4584
3/16	611	815	1019	1222	1426	1630	1833	2037	2241	2445	2648	2852	3056
1/4	458	611	764	917	1070	1222	1375	1528	1681	1834	1986	2139	2292
5/16	367	489	611	733	856	978	1100	1222	1345	1467	1589	1711	1833
3/8	306	408	509	611	713	815	917	1019	1120	1222	1324	1425	1528
7/16	262	349	437	524	611	698	786	873	960	1048	1135	1224	1310
1/2	229	306	382	459	535	611	688	764	840	917	993	1070	1146
5/8	183	245	306	367	428	489	550	611	672	733	794	857	917
3/4	153	203	254	306	357	407	458	509	560	611	662	713	764
7/8	131	175	219	262	306	349	393	436	480	524	568	613	655
1	115	153	191	229	267	306	344	382	420	458	497	535	573
1 1/8	102	136	170	204	238	272	306	340	373	407	441	476	509
1 1/4	92	122	153	183	214	244	275	306	336	367	397	428	458
1 3/8	83	111	139	167	194	222	250	278	306	333	361	389	417
1 1/2	76	102	127	153	178	204	229	255	280	306	331	357	382
1 5/8	70	94	117	141	165	188	212	235	259	282	306	329	353
1 3/4	65	87	109	131	153	175	196	218	240	262	284	306	327
1 7/8	61	81	102	122	143	163	183	204	224	244	265	285	306
2	57	76	95	115	134	153	172	191	210	229	248	267	287

point on the lip) can cause the cutting lips to chip. Therefore, if the twist drill is properly ground, decrease the feed to eliminate these conditions.

Precision drilling is required when two or more holes have been laid out in specific relationship to each other and the holes must be drilled accurately to layout marks. Templates, jigs, and fixtures are all used in precision drilling. This also includes drilling holes to a specified depth.

Drilling on Automatic Machines

High speeds and light feeds are especially recommended for automatic drilling machines where the holes do not exceed four drill diameters in depth. A small, compactly rolled chip is desirable. If possible, the chip should be kept unbroken for the entire depth of the hole, as such a chip feeds out through the flutes more easily. If the drill appears to be functioning well but the surface of the hole is rough, a dull twist drill is indicated, and it should be resharpened.

Controls

Most automatic drilling machines are controlled from a handy push-button control station. Coolant, spindle jog, spindle drill, and drill-tap operations are controlled by selector switches. Push buttons are used to control cycle start, spindle start, emergency spindle returns, and motor stop functions (Figure 4-22). The "jog" switch is provided for setups. The "on" position on the "jog" selector switch permits jogging the spindle downward—using the "cycle start" push button, or positioning the spindle with the handwheel on the lower part of the head. The "dwell" switch causes the spindle to pause at the bottom of the feed stroke for a period of time preset by the electric timing switch on the control panel. The "dwell" switch is noneffective in the tapping cycle. The "drill" position on the "drill-tap" selector switch causes the spindle to turn in the forward direction at all times. Set in the "tap" position, the spindle runs in the forward direction at the start of the cycle, reverses automatically when the bottom limit is reached, and then stops at the top of the travel (see Figure 4-22).

Depth of feed, upper limit of spindle travel, and start of power feed may be controlled by limit settings. Dial-type limit settings are used on some machines (Figure 4-23). The limit settings are quickly set on large concentric selector dials. The inner dial limits the return spindle stroke, the intermediate dial determines the disengagement

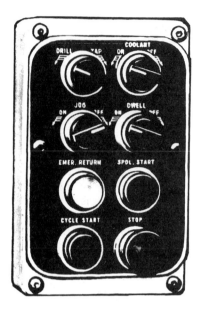

Figure 4-22 Control station on Bickford upright drilling machine, used to control all machine functions. *(Courtesy Giddings and Lewis Machine Tool Co.)*

of the rapid traverse and instantaneous engagement of the power feed, and the large outer dial determines the final depth of the tool.

Automatic Cycling

For automatic-cycle drilling, the spindle is advanced to the point set on the intermediate dial by depressing the "cycle start" button. The spindle advances at power feed to the depth setting determined by the outer dial. Then, rapid return is effected to the upper spindle position, determined by the inner dial, to complete the cycle (see Figure 4-23).

For automatic-cycle tapping, the spindle is rapidly advanced to the point set on the intermediate dial and continues at the tap lead rate to the depth determined by the outer dial. At the bottom limit, the spindle reverses, and the tap is retracted at the tap lead rate, until the tap is free of the work and returns to the upper limit at the rapid travel rate; spindle rotation stops, completing the cycle (see Figure 4-23).

Tapping

The drill press can be used as a means of *tapping* (or cutting threads in a drilled hole). Precision hand-tapping can be performed on the drill press with the work mounted on the drill press table in the position as

Figure 4-23 Dial-type limit-setting controls used on the Bickford upright drilling machine. *(Courtesy Giddings and Lewis Machine Tool Co.)*

for drilling the hole. A tap wrench is mounted on the square shank of the hand tap, and a lathe center is mounted in the hollow spindle of the drill press. Then, the tap is placed in the hole in the workpiece, and the point of the lathe center is placed in the center hole in the shank end of the tap. The hand-feed lever of the drill press is used to maintain steady pressure without forcing the hand tap, because the tap wrench is used to turn the tap into the work. Actually, the drill press is used, in this instance, chiefly as a guide for precision hand-tapping.

A tapping attachment can be used on the drill press. A friction clutch is built into the device. If a tap sticks, jams, or reaches the bottom of a blind hole, the clutch will slip before the tap will break. Raising the spindle of the machine permits a reversing mechanism in the tapping attachment to back out the tap without breaking it.

Automatic drilling machines can be equipped with one of three types of tapping devices:

- Friction-type tapping attachment
- Motor-reverse control operated by a small lever control
- Automatic reverse feed handle control

The friction-type tapping attachment is usually preferred to motor-reverse tapping for high-production work (Figure 4-24). Continuous motor reversals are not advisable for continuous operation on machines with back gears and power feed (Figure 4-25). Also, small taps, up to ⅜ inch, used for high-production work, perform better on the friction-type tapping attachment (Figure 4-26).

Countersinking

The operation in which a cone-shaped enlargement is formed at the end of a hole is called *countersinking*. A conical cutting or reaming tool is used to taper or bevel the end of the hole (Figure 4-27).

The countersinking operation involves fastening the workpiece properly, mounting the countersink in the drill chuck, aligning the countersink with the work, selecting the correct spindle speed, and using care in feeding the countersink to the work. The countersink should be rotated at a relatively slow spindle speed, using cutting oil for a smooth job in steel. Failure to clamp the work properly, a dull tool, or excessive spindle speed can be the cause of a rough countersinking job.

Counterboring

The operation in which a hole is enlarged cylindrically partway along its length is called *counterboring*. A counterboring tool is used to enlarge the diameter of the hole to accommodate a stud, bolt, or pin having two or more diameters—for example, a filister-head screw.

The counterboring tool (Figure 4-28) must be properly aligned with the original hole so that the pilot on the end of the counterbore will fit into and follow the original hole properly. The size of the enlarged hole is produced by the cutting edges on the counterbore.

Spot-Facing Operation

Another operation similar to counterboring is the *spot-facing operation* (Figure 4-29). Just enough metal is removed to provide a bearing surface for a washer, nut, or the head of a cap screw. The spot-facing tool should be mounted in the drill press spindle, and the pilot of the spot-facing tool should be aligned with the original drilled hole (Figure 4-30).

Figure 4-24 Older type of motor reversing mechanism used for taps ½ inch to 1 inch in diameter. A magnetic reversing control, actuated by a lever station mounted on the spindle arm, is used to reverse the rotation of the motor. *(Courtesy Buffalo Forge.)*

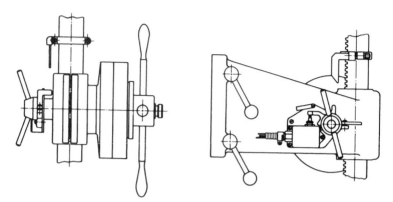

Figure 4-25 Automatic feed handle reverse tapping control (left) and power feed (right). *(Courtesy Buffalo Forge.)*

Figure 4-26 Using a Bickford radial drilling machine to tap a 5-inch bore in a motor housing. *(Courtesy Giddings and Lewis Machine Tool Co.)*

Figure 4-27 Machine countersink. The included angle is 82°.
(Courtesy Morse Twist Drill and Machine Co.)

Figure 4-28 Older model of straight-shanked machine-screw
counterbore. *(Courtesy Morse Twist Drill and Machine Co.)*

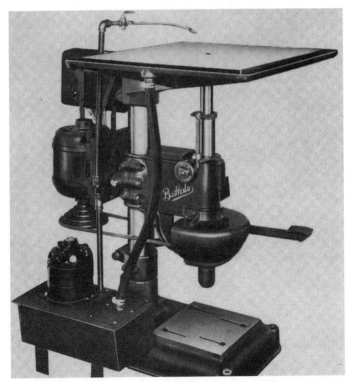

Figure 4-29 A spot-facing machine built to handle large forgings
and castings. This machine is equipped with a coolant system and
a foot feed. *(Courtesy Buffalo Forge.)*

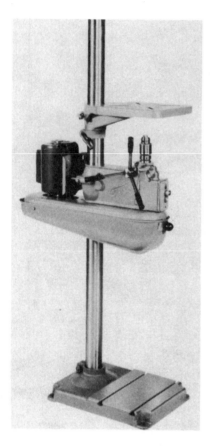

Figure 4-30 Inverted drill press head for routing or back spot-facing operations on the drill press. *(Courtesy Buffalo Forge.)*

Reaming

The *reaming* operation is used to accurately size and finish the inside of a hole. Reaming is used to make a drilled hole more accurate and smoother. A hand reamer can be used to finish a hole to the exact dimension. A hole should be drilled ¹⁄₆₄ inch undersize and then machine reamed to obtain an accurate smooth hole of standard size. If the blueprint should call for a ¹⁄₂-inch ream, a twist drill ¹⁄₆₄ inch under ¹⁄₂ inch should be used to drill the original hole.

Machine reaming can be performed in the drill press by substituting the correct size of reamer for the twist drill in the drill press spindle without changing the mounting position of the work or removing it. The correct spindle speed for machine reaming is approximately one-half the drilling speed. The automatic feed can be used for reaming.

The slight taper at the end of a reamer is provided to facilitate entry of the reamer into the hole, and the reamer should be extended through the hole a distance of at least 1½ inches to produce the accurate size. The reamer should never be rotated in the reverse direction, because the edges may be broken or dulled.

Grinding and Buffing

The drill press can be used for simple *grinding* and *buffing* operations. The work can be held in a drill press vise with a cup-grinding wheel mounted in the chuck.

A burring or polishing operation can be performed by mounting a buffing wheel in the chuck. The work should be held securely.

Other Operations

Many drill presses can be adapted for *mortising* operations by means of a mortising attachment (Figure 4-31). The attachment usually fits over the chuck, and the fence and hold-down are mounted on the table. A ½-inch-square hole is usually the maximum capacity for a mortising attachment on the drill press. *Trepanning* and *boring* operations can also be performed on the drilling machine (Figure 4-32).

Figure 4-31   Mortising attachment for a drill press.
(Courtesy Buffalo Forge.)

Holding Devices

Drill presses and drilling machines require two types of holding devices: devices for holding the tool or cutter, and devices that secure the workpiece in a properly mounted position for drilling, tapping, and so forth.

Tool or Cutter Holders

Most cutting tools have either a straight shank or a taper shank. Most drill presses are equipped with a Morse taper spindle. Straight-shank drills are held in a chuck.

Chucks

Many styles of chucks for a variety of purposes are manufactured (Figure 4-33 and Figure 4-34). The drill press chuck has a Morse

Figure 4-32 Using Bickford radial drilling machine for trepanning, boring, and tapping operations. Feed of 0.050 inch per revolution is used for boring and trepanning. Pipe tapping can be performed at 22 **rpm.** *(Courtesy Giddings and Lewis Machine Tool Co.)*

taper on the chuck shank. The tapered shank fits into the drill press spindle. Three small jaws in the chuck tighten simultaneously to hold the straight-shank twist drills, which are ½ inch and smaller in diameter.

Sleeves and Sockets

A tapered drill sleeve is used to hold taper-shank twist drills that are too small for the tapered hole in the spindle of the drill (Figure 4-35). The hole inside the sleeve is tapered to fit the tapered shank of a twist drill. The outside of the sleeve has a Morse taper to fit the hole in the spindle. The tanged end fits into the slot in the spindle.

A drill socket is used to hold twist drills with shanks too large to fit into either the drill press spindle or a sleeve (Figure 4-36). It also has a tanged end to fit into the slot in the drill press spindle.

A tapered key or *drill drift* is used to remove twist drills, sleeves, and sockets from the drill press spindle (Figure 4-37). These drifts

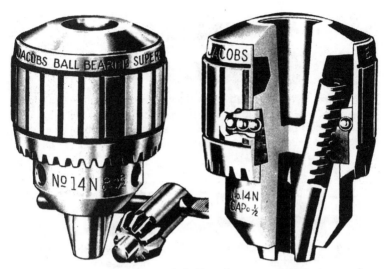

Figure 4-33 Ball-bearing chuck (left) and cutaway (right), mounted on either straight- or taper-shank arbors for use on drilling machines and other industrial equipment. *(Courtesy Jacobs Manufacturing Co.)*

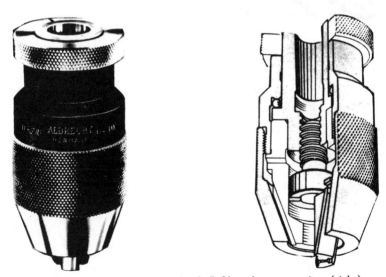

Figure 4-34 Albrecht keyless chuck (left) and cutaway view (right), used on high-speed sensitive drilling machines and other industrial equipment. *(Courtesy Jacobs Manufacturing Co.)*

Figure 4-35 Morse tapered nylon sleeve. The nylon sleeve is designed to give way under heavy loads before stress can ruin the drill. High resiliency helps prolong life of tools and equipment by absorbing deflections caused by loose spindles, excessive overhand, and dull tools. *(Courtesy Morse Twist Drill and Machine Co.)*

Figure 4-36 Steel socket used to hold twist drills with shanks too large to fit into either the drill press spindle or a sleeve. *(Courtesy National Twist Drill and Tool Co.)*

Figure 4-37 Drill drift used to remove twist drills, sleeves, and sockets from the drill press spindle. *(Courtesy National Twist Drill and Tool Co.)*

are made of tool steel and hardened. The drift is driven through the slot in a sleeve or socket to remove a taper-shank twist drill.

Work Holders and Setup Devices

The workpiece must be securely fastened to give satisfactory and safe results. Mounting and supporting the workpiece properly are extremely important. Of course, the layout work must be done properly, but the correct procedure for making setups and for operating the drilling machine must be followed carefully for satisfactory results.

Vise

A drill press vise is used to hold and support the work on the drill press table. A vise with a graduated swivel base is shown in Figure 4-38. An air-hydraulic vise is used on high-production machines (Figure 4-39). A combination table and vise for drilling machine operations is shown in Figure 4-40.

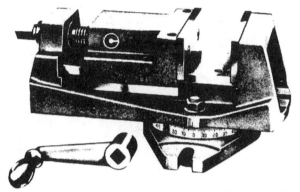

Figure 4-38 A Clausing machine vise with graduated swivel base.

(Courtesy Atlas Press Co.)

Figure 4-39 An air-hydraulic vise. *(Courtesy Wilton Tool Manufacturing Co., Inc.)*

Parallels

These accurately machined bars are made in pairs. *Parallels* are used to raise the work above the drill press table so that the part can be drilled completely through without damage to either the vise or the table. Parallels should be placed carefully so that the twist drill will not damage them after completing its passage through the work (Figure 4-41).

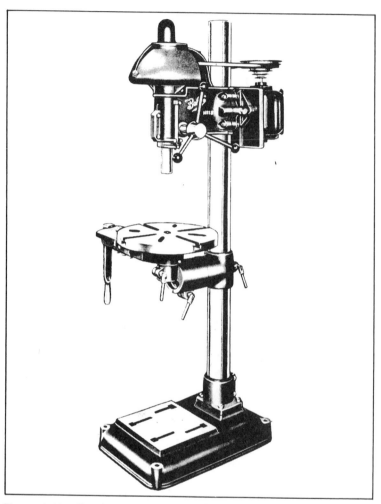

Figure 4-40 Combination table and vise for drilling machine operations. Note how the speed is changed manually. *(Courtesy Buffalo Forge.)*

V-Blocks

Round stock can be securely clamped with *V-blocks* for drilling (Figure 4-42). Care must be taken and the work clamped securely. Otherwise, the operator can be injured when the twist drill takes hold in the workpiece, causing the work to swing around.

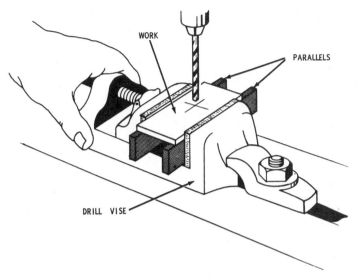

Figure 4-41 Using parallels in the drill press vise.

Figure 4-42 V-blocks and clamps made in pairs for use where extremely accurate settings are desired.
(Courtesy Lufkin Rule Co.)

Angle Plate
The *angle plate* is useful when it is desirable to drill a hole parallel to another surface. The angle plate is usually made of cast iron, and holes and slots are provided for clamping it to the machine table to

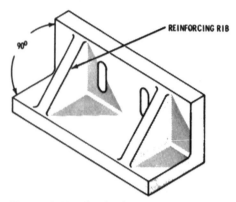

Figure 4-43   Angle plate.

secure the work. It is planed on two sides to an angle of exactly 90° (Figure 4-43).

T-Bolts

These bolts are placed in the T-slots provided in the table. Either the workpiece or the vise can be securely fastened to the table (Figure 4-44).

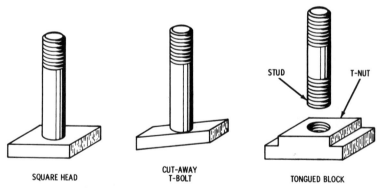

Figure 4-44   Common types of T-bolts: square head; cutaway T-bolt; tongued block.

Straps or Clamps

An assortment of straps or clamps can be used to clamp workpieces to the table. The clamps should be made of a good grade of steel to prevent bending under pressure (Figure 4-45).

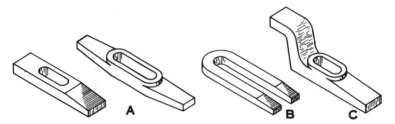

Figure 4-45 Common types of straps or clamps: (A) flat strap; (B) U-strap; (C) goose-neck strap.

Step Blocks
A *step block* is used to support the end of a clamp or strap opposite the work. These blocks are usually made of a good grade of steel and are usually made in pairs (Figure 4-46). A step block is useful as an aid to keeping the strap level when fastened to the work.

Safety
Any movement of either the workpiece or the table can cause the twist drill to break and result in injury to the operator. The

Figure 4-46 Step block.

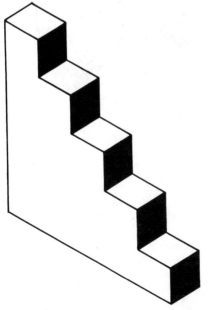

workpiece must be clamped securely and rigidly (Figure 4-47). Bolts should be just long enough and should be placed as near the work as possible. Clamps and blocks should be of correct height and size, and washers should be used on all nuts. Many injuries have been caused by attempting to hold the work with the hands. The operator should always make certain that the work is clamped securely.

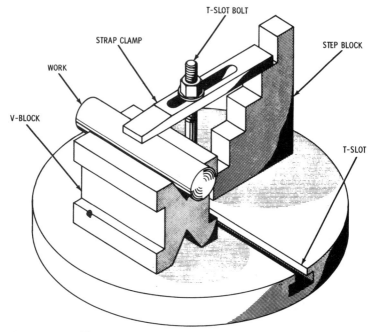

Figure 4-47 Typical setup in preparation for drilling in round stock.

Summary

The chief operation performed on the drill press is drilling. Drilling is the removal of solid metal to form a circular hole. Prior to drilling a hole in metal, the hole is located by drawing two lines at right angles, after which a center punch is used to make an indentation for the drill point at the intersection of the two lines to aid in starting the drill.

The operator should check to make certain that the drill point has started properly before the drill has cut too deeply. It is too late to shift the twist drill point (without marring the workpiece) after the drill has begun to cut its full diameter. After the work is placed on the worktable of the drill press, the table should be adjusted to a convenient height for drilling.

Many drill presses can be adapted for mortising operations by means of a mortising attachment. The attachment usually fits over the chuck, and the fence and hold-down are mounted on the table. Trepanning and boring operations can also be performed on the drilling machine.

Review Questions

1. What type of operations can be performed on a drill press?
2. What is the purpose of a center punch?
3. Name the parts of a twist drill.
4. What are the decimal and metric sizes of a No. 78 drill bit?
5. What is the decimal and metric equivalent of an "A" drill bit?
6. Why is speed important in a drilling operation?
7. What are the results of a drill bit with unequal lips?
8. High speeds and light feeds are especially recommended for _____ drill diameters in depth.
9. Most automatic drilling machines are controlled from a handy push-button _____ station.
10. The drill press can be used as a means of _____ or cutting threads in a drilled hole.
11. Name the three types of tapping services for automatic drilling machines.
12. What is an inverted drill press used for?
13. What is counterboring?
14. What is spot facing?
15. What is mortising?
16. What is a chuck?
17. What is a drill drift?
18. Identify the following: vise, parallels, V-blocks, angle plate, T-bolts, step blocks.
19. What are straps and clamps used for in drilling?
20. Why should you *not* hold work by hand when drilling?

Chapter 5

Vertical Boring Mills and Horizontal Boring Machines

The vertical boring mill is virtually equivalent to a lathe turned on end so that the faceplate is horizontal, omitting the tailstock. The table of the boring mill corresponds to the faceplate of the lathe. It is more convenient to clamp heavy work to the horizontal table of the mill than to the vertical faceplate of the lathe.

Although the vertical boring machine is adapted to boring and turning operations on work that has a diameter or width *greater* in proportion to its length, the horizontal boring machine is adapted to drilling, boring, and machining operations on work that has a diameter or width *smaller* in proportion to its length.

Vertical Boring Mill

The machine is adapted to boring and turning heavy work that has a larger diameter than its width or height. As on the lathe, the work turns—the only movement of the tool being caused by its feed. More than one tool can be used at a time. The feed is such that the cutting tool can be given horizontal, vertical, or angular movement. Thus, the work can be faced parallel with the table, bored, or turned cylindrically or conically.

The work can be either held in a chuck or clamped directly to the table. If the work has an irregular shape or size, clamping it to the table is more desirable. If a piece is to be machined on both sides, the previously machined surface can be placed on the table for more accuracy. On production work, special fixtures can be used to set up the work more quickly.

Classification

Boring mills are used for various purposes—drilling, reaming, chamfering, counterboring, and so on. They can be classified by the number of rams (either *single ram* or *double ram*), and they are available as either *single-* or *double-station* units (Figure 5-1).

Vertical boring mills are easy to load, especially for handling awkward parts. All stations are readily accessible front stations, making it easy for one operator to handle several machines. The

Figure 5-1 A vertical boring machine. *(Courtesy Heald Machine Co.)*

vertical mills also save space because their bulk extends upward into unused space, and they can be lined up close to each other.

The size of a boring mill is designated by the size of the table. For example, a 72-inch mill has a table that is 72 inches in diameter. As for the engine lathe, the size designation is called the *swing*.

Basic Construction

The basic units of the boring mill are the slide and the boring head. These components can be arranged in almost limitless combinations to provide efficient production line operation.

Slide units in combination with standard wing bases are designed to offer wide adaptability. They are rigid and stable, regardless of slide position (Figure 5-2). Slides can be mounted on either the right- or left-hand side of a transfer line, or as multiple operation units.

Figure 5-2 Slide unit for a vertical boring mill. *(Courtesy Heald Machine Co.)*

The boring head (Figure 5-3) is designed for rotating and feeding. It can handle almost any drilling, reaming, chamfering, and counterboring operation either as a single unit or in multiple spindle units. A number of the units can be grouped on a single base or on separate bases in an automated line. They can be mounted horizontally, vertically, or at any angle in any plane.

Figure 5-3   Boring head unit for a boring mill can be used in either single or multiple units. *(Courtesy Heald Machine Co.)*

The table of the boring mill (Figure 5-4) travels on box-type ways. Cross slides, steady rests, or workholding fixtures can be positioned on the table. The table shown in Figure 5-4 has a constant-feed hydraulic system of the locked-feed type to maintain constant feed rates.

Boring Mill Operations

The vertical boring mill operates on the principle that "it is easier to lay down a piece of work than it is to hold it upright," which is probably the chief reason for its development. Three general methods of fastening the work to the table are as follows:

- Bolts and clamps
- Chucks
- Fixtures

Boring

Holes in castings are usually cored, so it is necessary only to finish the rough hole to the required diameter. Usually, a boring tool is used for

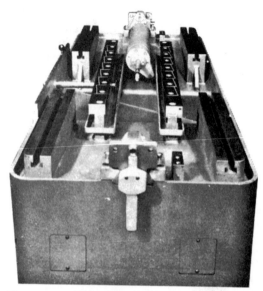

Figure 5-4 The table of the boring mill is mounted on hardened and ground box-type ways with hold-downs. Table travel is accurate even under heavy cuts. *(Courtesy Heald Machine Co.)*

small holes. Its shank is held in either a ram or a turret. Sometimes a shell drill with four flutes can be used. A very light finishing cut can be taken with a reamer for precision work (Figure 5-5).

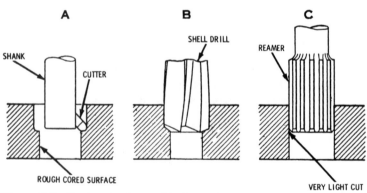

Figure 5-5 Boring small-cored holes with a shank cutter (A), with a shell drill (B), and finishing with a reamer (C).

For large holes, the tool can be held in a tool head or holder on the end of a ram (Figure 5-6). Several types of boring tools are available. The boring tool can be of a horizontal type or a bent type (see Figure 5-6). Cast iron is usually finished with a broad, flat tool.

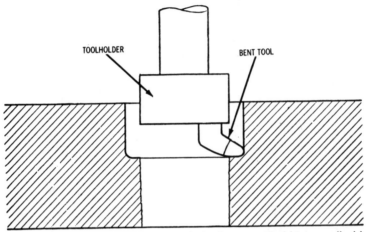

Figure 5-6   Boring a large hole with the bent tool held in a toolholder.

Holding the Work

Properly setting and fastening the work on the table is important in any boring mill operation. If the work to be bored is an irregular piece, it can be fastened to the table by the bolt and clamp method (Figure 5-7 and Figure 5-8).

The large hole in the work must be centered accurately with the center hole in the table. Because the hole to be bored is larger than the hole in the table, the work must be elevated above the table surface to provide clearance for the cutting tool at the lower end of the bore. This can be accomplished by means of parallel blocks (see A and B in Figure 5-8), but distance sleeves threaded on the bolts (see E and F in Figure 5-8) are preferable.

The upper surface of the work should be parallel with the surface of the table. This can be adjusted in several positions by making adjustments with the parallel blocks (see C and D in Figure 5-8). As the workpiece has a hole at each end, it can easily be fastened to the table by the bolts E and F. The bolts are placed in the T-slots of the table and passed through the end holes in the work, and the nuts are turned down firmly onto washers resting on the

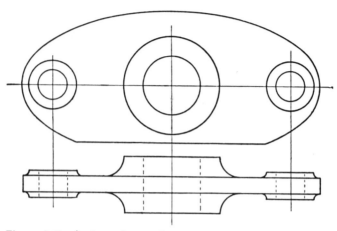

Figure 5-7 An irregular workpiece to be fastened to the worktable of the boring mill.

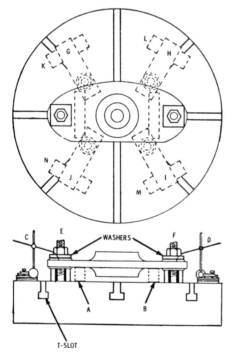

Figure 5-8 Using bolts and clamps to clamp an irregular workpiece to the worktable.

work. The concentric distance sleeves are preferable to the offset parallel blocks.

To determine whether the hole has been centered properly, fasten a scriber to the toolholder with its point near the circumference of the hole to be bored. Rotate the table slowly, and the scriber will indicate whether the hole has been centered properly for the operation.

If there are no holes at the end of the work, it can be fastened to the table with bolts and clamps (see G, H, I, J in Figure 5-8). The outer end of the flat plate clamps rests on the blocks (see K, L, M, N in Figure 5-8) at the proper height, as shown in detail in Figure 5-9.

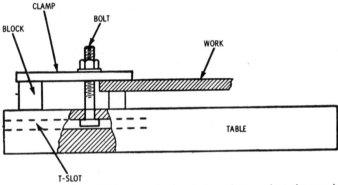

Figure 5-9 Detail drawing of a workpiece fastened to the worktable by means of bolts and clamps.

Turning
Although the machine is called a *vertical boring mill*, it can be used more for turning than for boring operations, especially for workpieces that require both turning and boring. The kinds of turning operations that can be performed on the boring mill are flat (facing), cylindrical, and conical.

Flat and Cylindrical Turning
A flywheel can be machined on the boring mill to illustrate both flat and cylindrical turning. Fastening the work to the table by chucking can also be illustrated.

Chucks that are built into the table are usually of the combination type (that is, the jaws have both universal and independent adjustment). The chuck jaws should be set against the interior surface of the rim, because the rim can be faced and turned at one setting.

To avoid the risk of the flywheel slipping in the chuck because of the tangential thrust of the cutting tool, place the wheel in the chuck, if possible, so that one spoke will bear against one of the chuck jaws. If this is not feasible, use a driver bolted to one of the T-slots. The driver can be an angle plate or any device that will prevent slippage in the chuck jaws. If a multistep inside the chuck jaws is used, the wheel can be clamped on the lower step to permit enough clearance above the table for the tool to turn the entire face without resetting (Figure 5-10 and Figure 5-11).

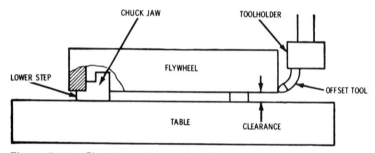

Figure 5-10 Chucking a flywheel inside the rim. Also, machining by cylindrical turning on a boring mill.

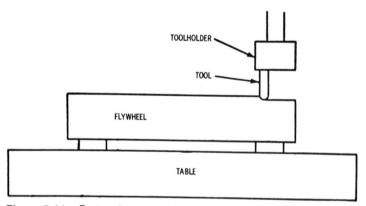

Figure 5-11 Facing the side of the rim of a flywheel. This shows flat turning on a boring mill.

After the tool has faced the rim, it is moved over to face the hub. Then the hole for the shaft is bored, and the flywheel can be turned over so that the other side of the rim and hub can be faced. The chuck jaws can be removed and the finished side of the rim placed on the table to ensure parallelism for the last operation. After the flywheel has been turned over, it can be centered by means of a plug inserted in the hole of the table. The upper section of the plug should be of proper size to fit the hole in the flywheel, which is clamped against the spokes.

Conical Turning

Usually, the saddle that carries the ram is arranged to swivel on a central stud by turning the angular adjustment gear, which consists of a worm and segment. Provision is usually made for swiveling the ram to 45° on either side of the vertical; the amount of swiveling is indicated on a graduated scale. The worm and segment also act as a lock to prevent tipping and to permit turning the ram to any angle within its range. If a desired taper or conical surface is given in degrees, the ram can be set by means of a scale.

To turn a taper or a conical surface, swivel the ram to the given angle, and use the vertical feed. A setup for turning a gasoline engine flywheel having an inside conical surface for a cone clutch is shown in Figure 5-12. The flywheel is held by a chuck and should be centered accurately, and the ram should be swiveled to the correct angle for the conical surface. The vertical feed should be set to

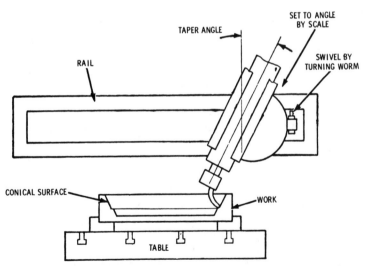

Figure 5-12 Machining the flywheel of a gasoline engine. This shows conical turning on a boring mill.

feed downward, starting the cut from the top. Take a first cut that will leave a smooth surface for checking with a taper gauge.

The taper gauge should be placed across the center and held square with the work. If the taper is correct, light cannot be seen at any point between the gauge and the conical surface. Take a second roughing cut, and check again with the taper gauge. If the angle is incorrect, the angular adjustment setting must be changed.

After the correct setting has been obtained, take the first finishing cut, leaving about 0.010 inch of metal for the last finishing cut. A finishing tool should be used for the last cut, and precautions should be taken not to remove too much metal. It is a good practice to take several light finishing cuts, checking with the gauge after each cut until the correct size is obtained.

Conical Turning by Combination of Vertical and Horizontal Feeds

If the angle of a conical surface is greater than the angular range of the ram, another means of machining must be used. Simultaneous use of both the vertical and the horizontal feeds can be made to achieve the correct angle. Angular adjustment of the head is necessary; the proper setting involves a trigonometrical calculation.

According to F. D. Jones, the calculation can be made as follows:

1. Suppose a conical casting is to be turned to an angle of 30° (Figure 5-13). The toolhead of the boring mill moves horizontally at 0.25 inch per revolution of the feed screw, and it has a vertical movement of 0.1875 (which is $\frac{3}{16}$) inch per turn of the upper feed shaft.

2. If the two feeds are used simultaneously, the tool will move a distance h of, perhaps, 8 inches while moving downward a distance v of 6 inches, thus turning the surface to an angle y. This angle is greater (as measured from a horizontal plane) than the angle required. However, if the tool bar is swiveled to an angle x, the tool (as it moves downward) will also be advanced horizontally, in addition to the regular horizontal movement.

3. As a result, the angle y is diminished, and if the tool bar is set over the correct amount, the conical surface can be turned to an angle a of 30°. Then, the problem is to determine the angle x for turning to a given angle a.

The calculation for the angle x is explained in connection with the diagram in Figure 5-14, which shows half the casting. Use a calculator to obtain accurate results.

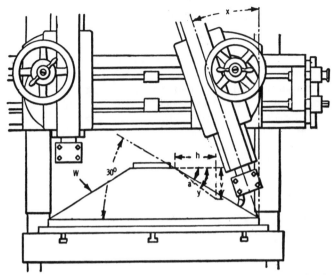

Figure 5-13 Combination of the vertical and horizontal feed movements in turning a conical surface on the boring mill.

1. The sine of the known angle *a* can be found in a table of natural sines. The sine of angle *b*, between the taper surface and the centerline of the toolhead, can be determined by the following equation:

$$\sin b = \frac{\sin a \times h}{v}$$

in which *h* represents the rate of the horizontal feed and *v* represents the rate of the vertical feed.

2. The angle corresponding to sine *b* can then be found in a table of sines. Thus, both angles *b* and *a* are known, and by subtracting the sum of these angles from 90°, the desired angle *x* can be obtained as follows:

   a. sine of 30° = 0.500

   b. sine of $b = \dfrac{0.500 \times 0.25}{0.1875} = 0.666$

   c. angle $b = 41.759°$, and angle $x = 90° - (30° + 41.759°) = 18.241°$

Therefore, to turn the casting to angle *a* in a boring mill having the horizontal and vertical feeds given, the toolhead should be set over 18.241° from the vertical.

3. If the required angle *a* were greater than angle *y*, obtained from the combined feeds with the tool bar in a vertical position, the lower end of the tool bar would swing to the left rather than to the right of the vertical plane. When the required angle *a* exceeds angle *y*, the sum of the angles *a* and *b* is greater than 90°, so that the angle *x* for the toolhead is equal to $(a + b) = 90°$.

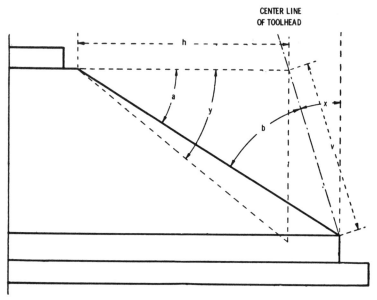

Figure 5-14   Diagram showing a method of obtaining the angular position of the toolhead for turning a conical surface by combining the vertical and horizontal feed movements of the boring mill.

Horizontal Boring Machine

The horizontal boring machine is very efficient for certain kinds of operations because nearly all the machining operations can be completed with just one setting on the machine (Figure 5-15).

Classification

Horizontal boring machines can be classified according to their method of holding the work (that is, either table-type or floor-type

Figure 5-15 A horizontal boring machine. These machines can be used to machine large heavy work, such as cylinder blocks and large pump or compressor bodies. *(Courtesy Heald Machine Co.)*

machines). The boring head can be either of the stationary or vertically adjustable type. The table can be either adjustable vertically or nonadjustable vertically. Some special types of horizontal boring machines are used for machining railway motors, locomotive engine cylinders, and so on.

Horizontal boring machines are rated according to size by the manufacturers, and each manufacturer can use its own method of rating. Generally, the size of the machine is given as the largest bar that the machine is designed to handle, but this may not be a true indication of the size of piece that the machine can handle for machining.

Basic Construction
Because most horizontal boring machines are designed to machine large pieces of work, they usually have the various parts assembled on a heavy, substantial bed. The head is mounted on a vertical column or post at one end of the bed.

Vertically Adjustable Head
The head post provides rigid support for the head in all operating positions. The entire head can be moved vertically on the ways of the column or head post (Figure 5-16). As the vertical position of the head is changed, the outboard bearing also moves up or down; the two parts are connected by shafts and gearing.

The head contains a sleeve in which a boring bar can move longitudinally. The bar carries cutters for boring operations, and either milling cutters or an auxiliary facing arm can be bolted to the end of the sleeve.

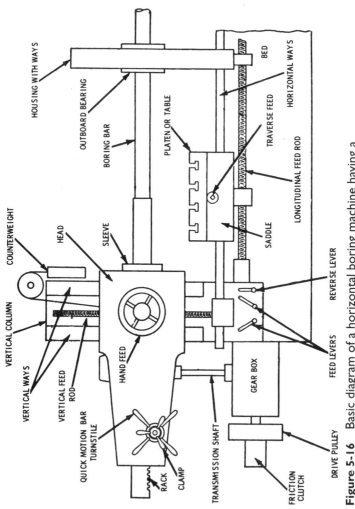

Figure 5-16 Basic diagram of a horizontal boring machine having a head that is adjustable vertically.

The end of the boring bar is held in alignment by the outboard bearing. The bar can be moved in either direction by hand-feed or by power feed. The entire head can be moved vertically on the column ways either by hand (to set the bar at the proper height) or by power (to feed a milling cutter in a vertical direction).

A saddle is mounted on the bed ways, and a table is mounted on the saddle (Figure 5-17). The saddle can move longitudinally and the table has a transverse feed, so the work attached to the table can be moved to any position for machining (that is, longitudinally, transversely, and vertically).

Figure 5-17   The table is mounted on the saddle, and the saddle is mounted on the bed ways of the horizontal boring machine.
(Courtesy Bullard Co.)

In actual operation of the boring machine, power is transmitted through the speed change gearbox and vertical transmission shaft to the head, and finally to the boring bar, causing it to rotate. The operator has a choice of many speeds for the work. Either hand-feed or power feed can be used for all the different feed movements.

Vertically Adjustable Table

In this type of machine, the table, rather than the head, is designed to move vertically (Figure 5-18). In the older machines, the spindle is cone-driven, and back gears are provided in the same manner as on a lathe.

The head is fixed and cannot be moved vertically, but the spindle can be moved longitudinally in the cone by means of a pinion that meshes with a rack that traverses the spindle. This movement can be effected by either a hand-feed or a power-feed mechanism.

Because the spindle cannot be adjusted vertically, the work can be adjusted for height by raising or lowering the table. This can be accomplished by means of worm-driven elevating screws. The worms that move the elevating screws are attached to the worm

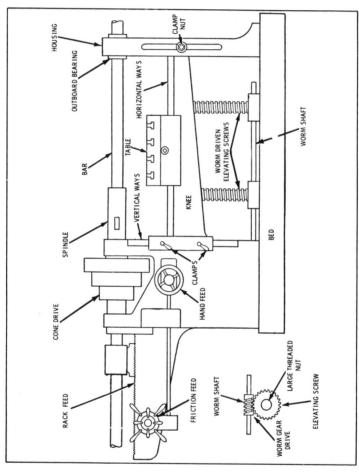

Figure 5-18 Basic diagram of a horizontal boring machine having a table that is adjustable vertically. Detail of the worm gear drive is shown in the lower left.

shaft; hence, each elevating screw is moved an equal amount. The knee, which carries the saddle and table assembly, traverses the vertical ways and is held firmly in any desired position by clamps at one end and clamp nuts at the other end.

Horizontal Boring Machine Operations

In the following methods of machining, examples of drilling, boring, and facing operations illustrate machining methods that are used on both the vertically adjustable spindle and the nonvertically adjustable spindle machines.

Drilling (Without a Jig)

A length of cast iron pipe with flanges at both ends is used to illustrate drilling, facing, and boring operations that can be performed on the horizontal boring machine (Figure 5-19). The dimensions, diameters of the bolt circles, and diameters of the holes are necessary for machining the piece.

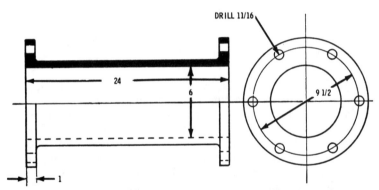

Figure 5-19 Working drawing of a length of flanged pipe used to illustrate machining on a horizontal boring machine with a head that is vertically adjustable.

The bolt circles on each flange must be laid out, marking the centers with a center punch. The work is mounted on V-blocks in the approximate position on the table of the machine, and clamped lightly (Figure 5-20). The position of the casting should be shifted so that two bolt-hole centers are at the same elevation above the table (that is, their scribed axis is horizontal, as determined by means of a surface gauge). Then, the casting must be aligned with the centerline, or spindle axis, of the machine.

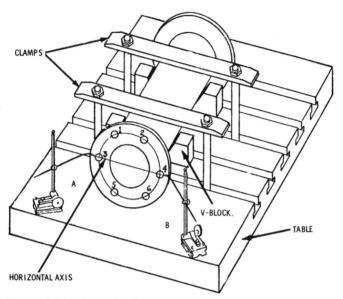

Figure 5-20 Setup for fastening and aligning the flanged pipe on the table of the horizontal boring machine.

The boring bar should be passed through the casting and mounted in the spindle and outboard bearing. The height of the boring bar should be adjusted so that it is centered with the bore vertically. With the calipers, measure the distance from the bore to the boring bar at the horizontal axis, and compare with the distance at the other end of the bar. Shift the work until the bore is equidistant from the bar at both ends. Recheck the horizontal axis with the surface gauge, and tighten the clamp bolts firmly. The horizontal axis should be trued, because two bolt holes can then be drilled without changing the setting of the vertical adjustment of the spindle.

Remove the boring bar, and place a drill of the proper size in the spindle. Use the cross-feed screw to adjust the spindle vertically to hole No. 1 (see Figure 5-20), and center the drill with the center-punch mark. Start the machine and cut a trial circle. Back off the drill and check to be certain that the work is centered properly with the drill. Finish drilling the hole (using the spindle feed); then use the cross feed to shift the work, and drill hole No. 2. To drill holes No. 3 and No. 4, lower the spindle to the correct height and drill as before. A third vertical adjustment can be made to drill holes No. 5 and No. 6.

Thus, the six holes can be drilled with only three vertical settings of the spindle. If the axis of the two bolt holes is not exactly horizontal, a vertical setting for each hole is necessary.

Each hole should be spot-faced on the back of the flange so that the nuts can bear smoothly on the flange when they are tightened on the bolts. Remove the drill and insert a spot-facing bar into the spindle. Place the spot-facing bar in the last drilled hole, and insert a fly cutter having a cutting length twice the diameter of the hole (Figure 5-21). Start the machine and, using slow feed, true or spot-face the surface of the flange around the hole. Repeat the operation for the other holes.

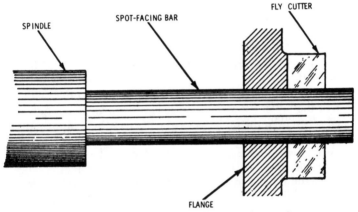

Figure 5-21   Detail drawing of a spot-facing bar with fly cutter.

Drilling (With a Jig)

In production work where many pieces are to be machined, a jig can save considerable expense because of the savings in time. Also, layout for each piece is avoided, and holes can be located and drilled with precision. The machined pieces are interchangeable and do not have to be fitted when a jig is used to machine them.

To construct a jig for the flanged piece, the jig must be laid out from the given dimensions (see Figure 5-19). When finished, the jig layout will appear as illustrated in Figure 5-22. Scribe a horizontal axis AB onto the jig and through the centers of the two opposite bolt holes, and continue the line BC across the edge of the jig. Scribe the axis AB and DE on both flanges of the flanged pipe, and continue across the edges of the flange. The axis must be identical in its position on both flanges.

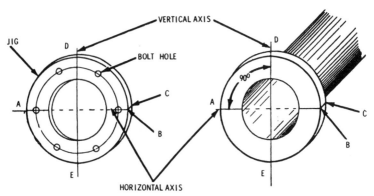

Figure 5-22 Jig (left) and flange layout (right) for machining the flanged pipe.

Mount the flanged pipe on the table of the horizontal boring machine, in the same manner as instructed for drilling without a jig, making certain that the axis *AB* is horizontal or parallel with the surface of the table. After the work has been aligned carefully and clamped securely to the table, place the jig over the flange that is to be bored, so that the scribed lines on the edge of the jig register with the scribed lines on the edge of the pipe flange. Clamp the jig to the pipe flange (Figure 5-23). If the jig should shift while it is being clamped, it should be driven back into position with a lead hammer. In drilling, use the spindle feed. Perform the boring operation in the same manner as described previously for drilling without a jig.

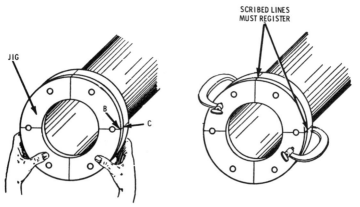

Figure 5-23 Positioning the jig (left) and clamping the jig to the flange (right).

Boring

After all the bolt holes have been drilled, boring and facing operations can be illustrated, using the same flanged pipe (Figure 5-24 and Figure 5-25). Without unclamping the work (because it is already in alignment with the spindle axis), mount a boring bar in the machine and center it with respect to the pipe bore, using the vertical spindle feed and the horizontal table feed.

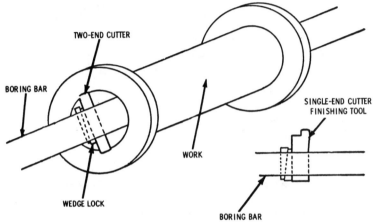

Figure 5-24 A boring bar with a two-end cutter position for machining. Detail of the single-end cutter used for finishing is shown (lower right).

A variety of boring tools are available (such as a boring head, and single-end and double-end cutters). The selection of boring tools depends on the nature of the work, dimension of the bore, amount of metal to be removed, and so on.

Move the table near the tailstock, because the table feed must be used. Mount the double-end cutter in the boring bar, and move the spindle with the hand-feed until the cutter is placed at the beginning of the cut. Lock the spindle feed. The spindle feed should not be used for the boring operation, because the machine would bore out of true if the bar should become misaligned.

Take a roughing cut with a double-end cutter that is just the correct length to leave enough metal for the finishing cut (see Figure 5-24). Use a single-end cutter for the finishing cut. If more than one finishing cut is necessary, the final cut should be very light (removing about 0.005 inch of metal). Note the shape of the finishing cutter in the detail drawing in Figure 5-24.

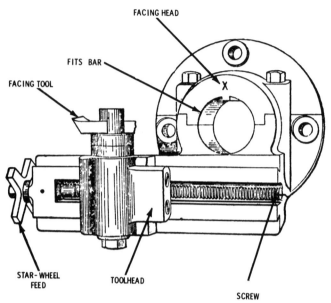

Figure 5-25 The facing head is rotated by the bar. The star wheel contacts a stop pin clamped on the bed of the machine. As a point of the star wheel strikes the stop pin, rotating the wheel, the screw that actuates the facing tool is, in turn, rotated.

Facing

In all turning machines, the facing operation is flat turning, in which the work is traversed with a cutting tool in a *plane* perpendicular to the axis of rotation. A facing head, as shown in Figure 5-25, can be used.

The setup for facing the pipe flange is shown in Figure 5-26. When mounting the facing head, place it in position to feed inward toward the bar. Clamp a stop pin onto the table in such a position as to engage one of the spokes of the star wheel on each revolution of the bar, and feed the cutter across the surface to be faced.

Take a cut across the flange, deep enough to remove the scale, and true up the flange. Then bring the tool back to its original position. Bring the work into position for the required depth of cut by means of the table feed, leaving enough metal for a light finishing cut. After finishing the first flange, transfer the facing head to the other end of the casting, and face the second flange in the same manner.

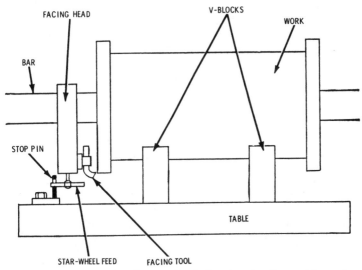

Figure 5-26   Facing head setup for facing the pipe flange. Use a slow cutting speed (not more than 35 feet per minute).

Summary

The vertical boring mill is virtually equivalent to a lathe turned on end so that the faceplate is horizontal, omitting the tailstock. The table of the boring mill corresponds to the faceplate of the lathe; it is more convenient to clamp heavy work to the horizontal table of the mill than to the vertical faceplate of the lathe. Boring mills are used for various purposes—drilling, reaming, chamfering, counterboring, and so on. They can be classified by the number of rams as either single ram or double ram, and they are available as either single- or double-station units. The size of the boring mill is designated by the size of the table. Three different ways of connecting the work to the table are bolts and clamps, chucks, and fixtures.

The horizontal boring machine is generally classified according to its method of holding the work as either a table or floor machine. Horizontal boring machines are rated according to size by manufacturers, and each manufacturer can use its own method of rating. Generally, the size of the machine is given as the largest bore that the machine is designed to handle.

Most horizontal boring machines are designed to machine large pieces of work; they usually have the various parts assembled on a heavy, substantial bed. The head is mounted on a vertical column or post at one end of the bed.

Review Questions

1. What makes the vertical boring mill different from the engine lathe?
2. What types of work can be performed on a vertical boring mill?
3. How is the size of a vertical boring mill determined?
4. What are the three methods of fastening the work to a boring mill table?
5. What is the main difference between a vertical boring machine and a horizontal boring machine?
6. How are horizontal boring machines classified?
7. What determines the size of a horizontal boring machine?
8. What are the basic units of the boring mill?
9. What are the three general methods of fastening work to the vertical boring mill table?
10. What type of tool is used to finish cast iron?
11. Although the machine is called a vertical boring mill, it can be used more for _____than for boring operations.
12. What is conical turning?
13. What is a saddle used for on a horizontal boring machine?
14. How do you drill without a jig?
15. In all turning machines, the facing operation is _____ turning.

Chapter 6

Lathes

The screw-cutting engine lathe is the oldest and most important of all the machine tools. Practically all of our modern machine tools have been developed from it.

Design and Functions

Although lathes have changed greatly in both design and appearance, the fundamental principles of design, construction, and operation have remained virtually the same (Figure 6-1). Therefore, a thorough understanding of the lathe (and the work that can be performed on it) enables a skilled worker to operate almost any lathe, regardless of its age or make.

The engine lathe produces cutting action by rotating the workpiece against the edge of the cutting tool. Lathe work is usually in the shape of a cylinder because the lathe produces a surface that is curved in one direction and is straight in one direction. The line of cut follows a curve (circumference of a circle), and the path of consecutive cuts produces a straight line to form the cylinder.

Size of Lathe

The maximum size of work that can be handled by the lathe is used to designate the size of the lathe (that is, the diameter and length of the work). Usually, the maximum diameter of the work is specified first, and the length is specified as the maximum distance between lathe centers (Figure 6-2).

Manufacturers usually use the term *swing* to designate the size as the maximum diameter of the work that can be machined in the lathe. Thus, a 16-inch lathe indicates that the machine is designed to machine work up to 16 inches in diameter. Manufacturers also use the *length of the lathe bed*, rather than the distance between centers, to indicate lathe size.

The swing of a lathe refers to the *nominal* swing and not to the actual swing of the lathe. The actual swing is the radial distance from the center axis to the bed and is always slightly greater than the nominal swing. This clearance margin can vary from ½ inch for a small lathe to 1½ inches for a large lathe to prevent any projection irregularities on the work coming in contact with the lathe and resulting in damage. The rated swing of a lathe refers to the *lathe*

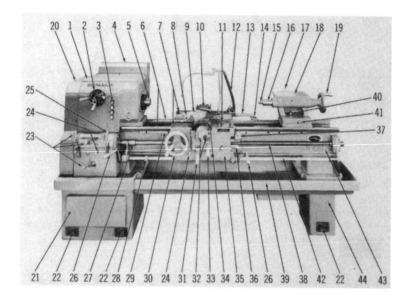

Figure 6-1   Functional diagram of an engine lathe. *(Courtesy Cincinnati Milacron Co.)*

1. Headstock and tray
2. Spindle speed selector dial
3. Push button panel for start, stop, reverse, and coolant pump switch
4. Electrical compartment
5. Headstock spindle and spindle nose
6. Ways
7. Saddle (a component of the carriage assembly)
8. Compound rest dial (direct reading) and adjusting handle
9. Compound rest (a component of the carriage assembly
10. Back tool rest (part of cross slide)
11. Carriage system
12. Coolant system
13. Taper attachment bracket
14. Taper attachment clamp
15. Dead center in tailstock
16. Tailstock spindle
17. Tailstock spindle clamp
18. Tailstock and tray
19. Tailstock handwheel
20. Guard or cover for motor drive and gear train
21. Storage compartment
22. Leveling screw
23. Quick-change gear box and levers for settling lead and feed gears.

24. Oil shot system
25. Reversing lever for lead and feed
26. Headstock start and stop lever (two levers)
27. Bracket and clutch stop for longitudinal feed
28. Stop clamp for longitudinal feed
29. Rack for longitudinal feed
30. Handwheel for longitudinal feed
31. Longitudinal power feed engaging lever
32. Dial indicator (direct reading) and adjusting handle for cross slide
33. Cross slide power feed engaging lever
34. Cross slide (a component of the carriage assembly)
35. Apron (a component of the carriage assembly)
36. Half-nut engaging lever for threading
37. Thread dial for chasing dial
38. Lead screw and feed rod—feed is run off a keyway in the lead screw
39. Chip man
40. Clamp nut for tailstock
41. Adjusting screw for tailstock alignment or offset
42. Start and stop bar or control rod
43. Bed
44. Legs or base

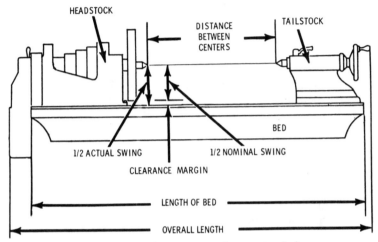

Figure 6-2 Note the size designations of an engine lathe.

bed and not to the *lathe carriage*. The swing over the carriage is necessarily less than the swing over the bed (Figure 6-3).

Types of Lathes
Engine lathes vary widely from the small bench lathe (or *tool-room lathe*) to the large *gap-bed lathes* and special-purpose lathes. Tool-room lathes are used to machine small parts. Larger and heavier

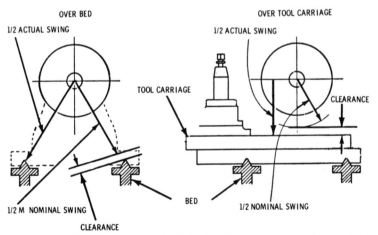

Figure 6-3 The swing over the lathe bed (left), and over the carriage (right).

lathes are used in high-production work. Examples of some of these high-production machines are turret lathes, multiple-tool lathes, and multiple-spindle lathes.

Basic Construction

A lathe is made up of many parts (see Figure 6-1). Following are the principal parts of a lathe:

- Bed
- Headstock
- Tailstock
- Carriage
- Feed mechanism
- Thread-cutting mechanism

Bed

The lathe *bed* (Figure 6-4) is a stationary part that serves as a strong, rigid foundation for a great many moving parts. Therefore, it must be scientifically designed and solidly constructed.

The *ways* of the lathe serve as a guide for the saddle of the carriage as it travels along the bed, guiding the cutting tool in a straight line. The *V-ways* (Figure 6-5) are machined in the surface of the bed, and are precision-finished to ensure proper alignment of all working parts mounted on the bed.

The two outer V-ways guide the lathe carriage. The inner V-ways and the flat way together provide a permanent seat for the headstock and a perfectly aligned seat for the tailstock in any position. A slight twist in the bed of the lathe can cause the machine to produce imperfect work. The lathe should be carefully leveled in both lengthwise and crosswise directions.

Headstock

The *headstock* is mounted permanently on the bed of the engine lathe at the left-hand end of the machine. It is held in alignment by the ways of the bed and contains the gears that rotate the spindle and workpiece (Figure 6-6).

Controlling Headstock Spindle Speed

On a belt-driven *back-geared lathe*, the different steps of the cone pulley and the back gears are used to make changes in spindle speed. On the three-step pulley, three speeds can be obtained by driving directly from the pulley to the spindle without the back gears in

Figure 6-4 Lathe bed and ways of an engine lathe.

(Courtesy Cincinnati Milacron Co.)

mesh. The smallest step gives the fastest speed. Three slower speeds can be obtained by driving the spindle through the back gears. Thus, three speed changes in direct drive and three speeds in back-gear drive permit a total of six changes of spindle speed.

The back gears can be meshed as follows:

1. Stop the machine.

2. Disengage the locking pin (the spindle will not move while the locking pin is disengaged, but the pulley will run freely).

3. Engage the back gears by pulling the handle forward to mesh the gears.

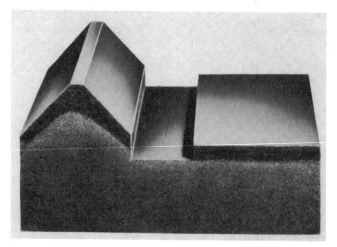

Figure 6-5   This is one of the V-ways and flat ways of a lathe bed. *(Courtesy Cincinnati Milacron Co.)*

Figure 6-6   A lathe gearing mechanism enclosed in the headstock housing. *(Courtesy Cincinnati Milacron Co.)*

Various types of controls are used on the *geared-head* lathes (Figure 6-7). Usually, selector levers are used, and the entire range of spindle speeds is obtained by reading directly from the dial.

Spindle
The hollow *spindle* is built into the headstock with the *spindle nose* projecting from the housing of the headstock. The spindle is hollow throughout, and is tapered at the nose end to receive the live center.

Figure 6-7 Spindle-speed selector on the headstock of a gear-head lathe. A "color match" dial is used to select the desired **rpm.** *(Courtesy Cincinnati Milacron Co.)*

The spindle nose is usually threaded. Either a faceplate or a chuck can be turned onto the threaded nose spindle to support and rotate the workpiece (Figure 6-8).

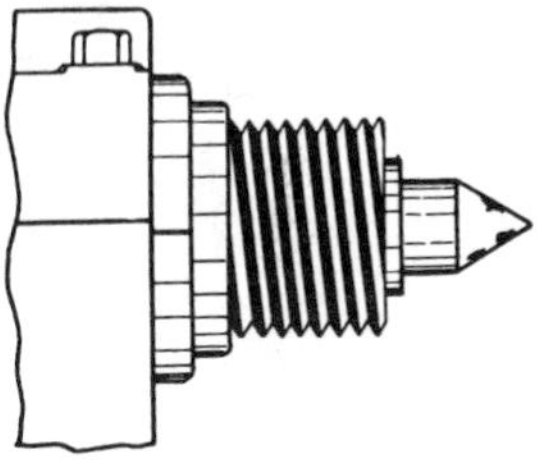

Driver Plate
The *driver plate* turns onto the threaded spindle nose. Its purpose is to drive the work mounted on the lathe centers by means of a *lathe dog*, which is clamped to the workpiece. The bent end of the dog engages a slot in the driver plate (Figure 6-9).

Figure 6-8 A threaded spindle nose with a live center. A faceplate, a driver plate, or a chuck can be turned on to the threaded spindle nose to support the workpiece and to rotate it.

Tailstock
The *tailstock* assembly is movable on the bed ways, and carries the tailstock spindle (Figure 6-10). The tailstock spindle has a standard Morse taper at the front end to receive a dead center. The tailstock handwheel is at the other end to give longitudinal movement when mounting the workpiece between centers. Reamers and taper-shank twist drills can be mounted in the tailstock spindle when required. A spindle binding lever clamps the spindle in any position in its travel. Clamp bolt nuts are used to clamp the tailstock assembly in any position of its travel on the ways of the bed. The dead center or any other tool mounted in the tailstock spindle can be removed by turning the tailstock handwheel counterclockwise.

Carriage
The carriage assembly is the entire unit that moves lengthwise along the ways between the headstock and the tailstock (Figure 6-11). The carriage supports the *cross slide*, the *compound rest*, and

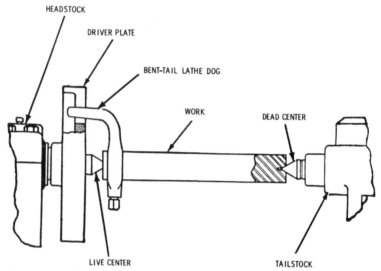

Figure 6-9   Driver plate and bent-tail lathe dog used to drive work mounted between the lathe centers.

Figure 6-10   Tailstock assembly. *(Courtesy Cincinnati Milacron Co.)*

Figure 6-11   The carriage assembly including saddle, cross slide, compound rest, and apron. *(Courtesy Cincinnati Milacron Co.)*

the *tool post*. The two main parts of the carriage are the *saddle* (Figure 6-12) and the *apron*.

Saddle
The saddle is an H-shaped casting that is machined to fit the outer ways of the lathe bed. The saddle can be moved along the ways either manually or by power, through the gearing mechanism in the apron. The apron and the cross slide are bolted to the saddle.

Cross Slide
The cross slide is a casting that is mounted on and gibbed to the saddle (see Figure 6-11). The cross-slide screw is located in the saddle and is connected to the cross slide. The cross slide can be moved either manually or by power across the saddle in a plane perpendicular to the longitudinal axis of the spindle.

The cross-feed micrometer dial is graduated in thousandths of an inch and is a very precise measuring instrument to be used when taking a cut on the work in the lathe. The operator must remember that when the dial is turned into the work one thousandth of an inch, the diameter of the workpiece is reduced by two thousandths of an inch because metal is removed from both sides of the continuously-turning cylindrical work.

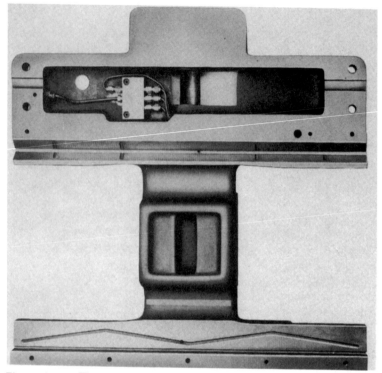

Figure 6-12   The saddle is the H-shaped part of the carriage assembly. It is machined to fit and is gibbed to the outer ways of the lathe bed.
(Courtesy Cincinnati Milacron Co.)

Compound Rest

The compound rest is mounted on the cross slide (see Figure 6-11). It consists of two main parts—the base and the slide. The base can be swiveled to any angle in a horizontal plane; the slide can be moved across the base by making hand adjustments with the micrometer dial, which is graduated in thousandths of an inch. The compound rest supports the cutting tool, and makes it possible to adjust the tool to various positions.

The *tool post* assembly is mounted on the compound rest (Figure 6-13). The rocker or wedge has a flat top, and is convex in shape on the bottom to fit into a concave ring or collar so that the cutting tool can be centered. A heavy-duty, four-way turret tool post is shown in Figure 6-14.

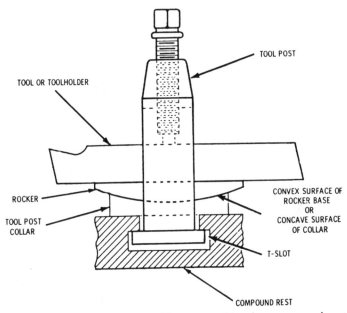

Figure 6-13 Tool-post assembly mounted on the compound rest.

Figure 6-14 A heavy-duty, four-way turret tool post.

(Courtesy Cincinnati Milacron Co.)

Apron

The apron is bolted to the front of the saddle. The apron houses the gears and controls for the carriage and the feed mechanism (Figure 6-15 and Figure 6-16). The carriage can be moved either manually or by engaging the power feeds.

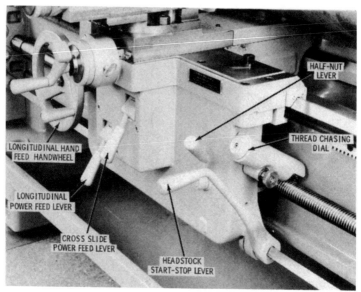

Figure 6-15 Front view of the apron, which is bolted to the saddle, on the front of the lathe. *(Courtesy Cincinnati Milacron Co.)*

The longitudinal feed handwheel, the longitudinal power-feed lever, the cross-slide power-feed lever, the half-nut lever, and the thread-chasing dial are all located on the apron (see Figure 6-15). The "Start-Stop" lever for the headstock spindle is usually located near the apron, convenient to the operator when he is standing in front of the apron.

Feed and Thread-Cutting Mechanism

The same gears that move the carriage are involved in the feed and thread-cutting mechanisms. These gears are used to transmit motion from the headstock spindle to the carriage.

Lead Screw

Some lathes have two lead screws—one for turning operations and one for thread-cutting operations exclusively. The lead screw is very strong and has coarse, accurate threads (see Figure 6-16).

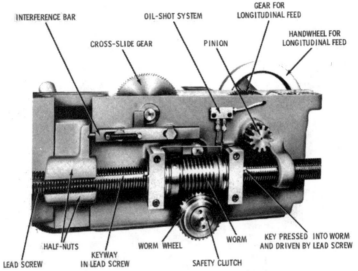

Figure 6-16   Apron internal gearing (rear view). *(Courtesy Cincinnati Milacron Co.)*

Quick-Change Gearbox

On most lathes, the quick-change gearbox is located directly below the headstock on the front of the lathe bed (Figure 6-17). A wide range of feeds and threads per inch (TPI) may be selected by positioning the gears. The *index plate* is an index to the lever settings required to position the gears for the different feeds and numbers of threads per inch. The distance, in thousandths of an inch, that the carriage will move per revolution of the spindle is given in each block for the corresponding gear setting.

The *reversing lever* is used to reverse the direction of rotation of the screw for chasing right- or left-hand threads, and for reversing the direction of feed of the carriage assembly. Levers on the quick-change gearbox should *never* be forced into position.

Holding and Driving the Work

A number of work-holding devices are used to hold or support the workpiece securely for the different lathe operations. A number of different devices are also used as toolholders.

Chucks

Some workpieces must be held and rotated by chucks—depending on the size, shape, and operation to be performed. The chucks are attached to the headstock spindle of the lathe.

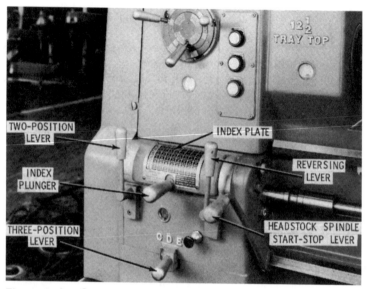

Figure 6-17 Quick-change gearbox. A wide range of feeds and threads per inch can be selected. *(Courtesy Cincinnati Milacron Co.)*

Great care should be exercised both in attaching and removing chucks from the headstock spindle. In attaching chucks to the *threaded spindle nose*, make certain that the spindle threads and the chuck threads are free from dirt and grit. The chuck turns onto the spindle in a clockwise direction—finish tightening the chuck with a quick spin by hand. Rotation of the workpiece against the cutting tool tightens the chuck on the spindle threads. A board or chuck block should be placed across the ways of the lathe to protect them when either mounting or removing the chuck.

Some lathes have a standard taper *key-drive spindle* (Figure 6-18). In mounting, the lathe spindle should be turned so that the key is facing upward. Both the lathe spindle and the chuck taper should be free of grit, oil, and chips, especially along the spindle key. In positioning the chuck on the spindle, make sure the keys are in alignment before tightening the draw-nut collar. Then, the draw-nut collar should be tightened securely.

Another type of spindle found on some lathes is the standard *camlock spindle* (Figure 6-19). Here again, great care must be exercised in mounting and in removal to ensure accuracy and prevent damage.

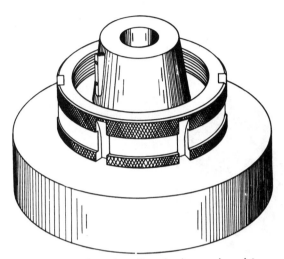

Figure 6-18 American Standard taper key-drive spindle. *(Courtesy Cincinnati Milacron Co.)*

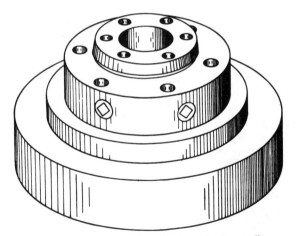

Figure 6-19 American Standard camlock spindle.
(Courtesy Cincinnati Milacron Co.)

Three-Jaw, or Universal, Chuck

All three jaws of the universal chuck are moved at the same time in the chuck body. This chuck is usually used for either round or hexagonal stock. Because the three jaws move simultaneously and are always the same distance from center, the workpiece is automatically centered without further adjustment (Figure 6-20). However, some accuracy may be lost because of wear. To ensure accuracy in machining, a workpiece should never be removed or reversed until all operations on the work have been completed.

Figure 6-20   Burned three-jaw, or universal, chuck.

(Courtesy Atlas Press Co.)

Four-Jaw, or Independent, Chuck

The four jaws are adjusted independently in this type of chuck, and the jaws are reversible so that work of any shape can be clamped from either the inside or the outside of the jaws. The independent chuck is one of the most accurate means of chucking a workpiece (Figure 6-21).

Figure 6-21   Burner four-jaw, or universal, chuck. (Courtesy Atlas Press Co.)

Mounting the work in a four-jaw chuck is largely a problem of centering. The concentric rings on the face of the chuck can be used as a guide in centering the workpiece. The work should be clamped as closely centered as possible. Then, test for trueness, marking the

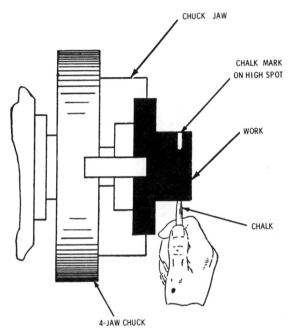

Figure 6-22 Using a piece of chalk to center the workplace in a four-jaw, or independent, chuck.

high spots with a piece of chalk resting against either the tool post or a tool mounted in the tool post (Figure 6-22).

Brass strips or shim stock with a piece of emery cloth between the shim stock and the workpiece should be used on work having a finished surface to prevent the chuck jaws cutting into the surface finish. This also allows the work to be moved in the jaws for alignment.

Pointers on the Use of Chucks

Proper use and care of chucks is important. Following are some pointers that should be observed in handling and caring for these precision tools:

- Keep the chuck clean; do not oil excessively—a light film of oil on all working parts is ample.
- Remove the live center and sleeve from the spindle before mounting the chuck.

- Clean the face of the shoulder on the spindle nose and the back face of the chuck.
- Place a few drops of oil on the threaded spindle nose.
- Clean the threads in the chuck and on the lathe spindle with a clean brush before mounting a chuck on a threaded spindle nose.
- When the chuck is close to the shoulder, give it a quick spin by hand to tighten it. A soft thud indicates a good firm seating against the shoulder. Forcing a chuck suddenly against the shoulder strains the spindle and makes chuck removal difficult. The chuck is kept tight by rotation while the lathe is in operation.
- Do not apply too much pressure while tightening the work in the chuck jaws. Excessive pressure may spring the work and affect the accuracy of the chuck.
- Tighten the jaws around the more solid parts of the workpiece; always use the chuck wrench made for the chuck.
- Turn the work as the jaws are tightened for a good fit.
- Small-diameter work should not project more than four or five times the workpiece diameter from the chuck jaws—cuts should be short and light.
- Heavy cutting pressures often cause the work to spring out and "ride the tool." Long work should be supported by the tailstock center, a steady rest, or a follower rest.
- Do not force the chuck to carry work larger in diameter than the diameter of the chuck body. Repeated overloading can damage the chuck.
- If the jaws stick, tap lightly with a piece of wood and move them in and out to remove chips—this indicates that the chuck should be taken apart for a thorough cleaning. An old toothbrush is excellent for cleaning a chuck. Wash the parts in kerosene. Do not use excessive oil in reassembling the chuck; oil collects dust and chips, which clog the chuck mechanism.
- Keep the chuck in a rack or bin when not in use. Dirt, dust, chips, and falling tools can cause much damage.
- Place a board across the top of the ways to protect them from damage when either removing or mounting a chuck.
- Never use an air hose to clean a chuck.
- Never leave the chuck wrench in the work.

Draw Collet Chuck

The *collet* is the holding or gripping part of the chuck. The assembly is comprised of a hollow draw-in spindle or bar, which extends through the headstock spindle. The split collet is held in a tapered holding sleeve (Figure 6-23).

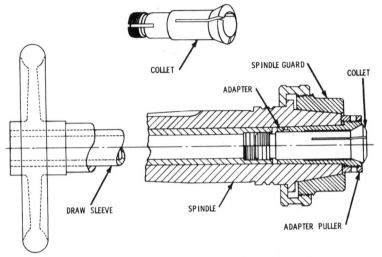

Figure 6-23 A draw collet chuck as mounted on the headstock spindle. *(Courtesy Cincinnati Milacron Co.)*

A collet chuck is used for precision work in making small parts. The workpiece held in the collet should be not more than 0.001 inch smaller or 0.001 inch larger than the nominal size of the collet. A different collet should be used for each work diameter. If the work diameter is not within the above limits of the collet size, the accuracy and efficiency of the collet is impaired. To prevent distortion, a collet should never be closed unless a suitable plug or piece of work is inserted in the hole of the collet before it is closed.

Collets are available for various shapes of stock—round, square, and so on. A new type of collet chuck and collet is shown in Figure 6-24.

Faceplates

The workpiece can be mounted or held on a faceplate, which is a round metal plate that can be mounted on the headstock spindle (Figure 6-25). The work is clamped or bolted to the elongated slots

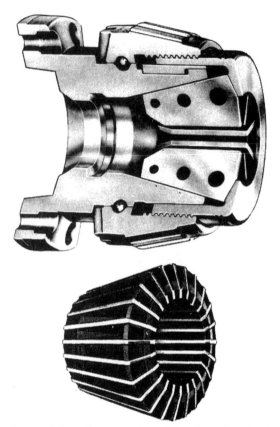

Figure 6-24 Cutaway view of Jacobs collet chuck (top), and showing a rubberflex collect collet (bottom). *(Courtesy Jacobs Manufacturing Co.)*

of the faceplate. The flat surface of the faceplate is perpendicular to the spindle.

Lathe Centers

The *live center* is stationary in the main spindle of the headstock. It is called "live" because it turns with the workpiece. The *dead center* is placed in the tailstock spindle, and it does not rotate with the workpiece (hence the term "dead" center).

The live center does not have to be hardened because it rotates with the work. The dead center must be hardened because it does not rotate and must withstand the friction of the workpiece against it. Hardened centers are usually indicated by grooves cut in the circumference.

Figure 6-25 Faceplate.
(Courtesy Cincinnati Milacron Co.)

Both centers usually have a standard Morse taper shank to fit in the tapered hole in the spindle, and a 60° cone end. The 60° angle may be turned on a lathe for the unhardened live centers; the dead or hardened centers are ground to the 60° angle (Figure 6-26).

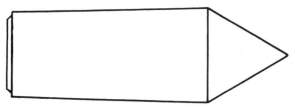

Figure 6-26 A standard lathe center with 60° cone end.
(Courtesy Cincinnati Milacron Co.)

Live centers are removed from the headstock spindle by placing a knock-out bar through the hollow spindle to bump out the center. Care must be exercised to keep the center from striking the ways of the lathe to avoid damage to either the center or the machine.

Some live centers are made for the machining of pipe and similar hollow-shaped work. The heads are constructed of hardened and ground steel to provide good resistance to abrasive wear (Figure 6-27). The bearings are protected by a two-piece seal, and accuracy is ±0.0001 inch. All pipe and bull heads come with a standard 60° included angle to match the chamfer on work for truer running.

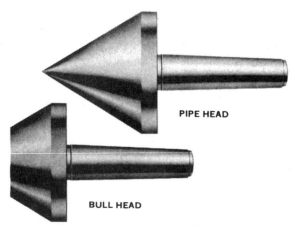

Figure 6-27 Pipe-head and bull-head live lathe centers.
(Courtesy Royal Products)

Several types of centers are available for various purposes. A *female center* is shown in Figure 6-28. The *ball center* (Figure 6-29) is used when the "tailstock setover" method of turning tapers is used. If the standard live center is used in turning tapers by this method, proper bearing on both the center and the center hole may not be provided, resulting in damage to both the center and the workpiece.

Figure 6-28 A lathe female center.
(Courtesy Cincinnati Milacron Co.)

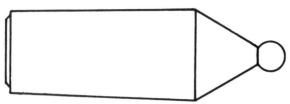

Figure 6-29 Ball center. *(Courtesy Cincinnati Milacron Co.)*

The *half center* (Figure 6-30) and the *crotch center* (Figure 6-31) are used only as tailstock centers. The half center is used in machining small-diameter workpieces and in facing operations where it is necessary to bring the facing tool near the center hole of the workpiece. The crotch center is used to hold a workpiece that is to be drilled at a location other than the ends. A revolving center (Figure 6-32) rotates with the workpiece to eliminate the friction of the workpiece on the center. It is used in the tailstock of the lathe.

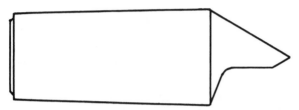

Figure 6-30 Half center. *(Courtesy Cincinnati Milacron Co.)*

Figure 6-31 Crotch center. *(Courtesy Cincinnati Milacron Co.)*

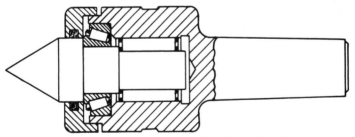

Figure 6-32 A revolving center used for the tailstock center.
(Courtesy Cincinnati Milacron Co.)

Lathe Dogs

The lathe dog is used with the drive plate as a simple means of causing the workpiece to rotate with the live center when the workpiece is mounted between centers (Figure 6-33). The *straight dog* and the

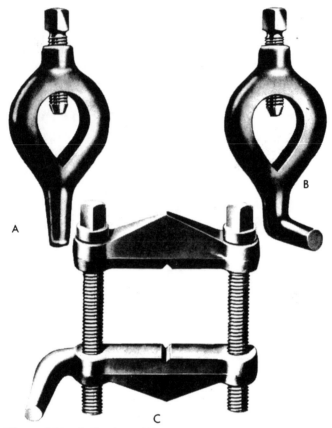

Figure 6-33 Lathe dogs: (A) straight dog; (B) bent-tail dog; (C) clamp dog. *(Courtesy Cincinnati Milacron Co.)*

bent-tail dog are used for round work, and the *clamp dog* is used for work having a flat side.

The bent-tail dog and the clamp dog have a tail piece that is inserted in a slot of the driver plate. Shim stock should be used as a collar for the workpiece to prevent damage by the dog.

Mandrels

The standard lathe mandrel is a shaft or bar with 60° centers so that it may be mounted between centers. A workpiece is mounted on the mandrel for turning its outside surface true with the center hole. The mandrel is always rotated with a lathe dog; it is never placed in a chuck for turning the workpiece.

A *straight lathe mandrel* (Figure 6-34) is designed to fit the entire length of the center hole in the workpiece. A *tapered mandrel* (Figure 6-35) is used with parts that have a standard taper center hole. Workpieces having two internal diameters can be turned on a *double-diameter mandrel* (Figure 6-36). The eccentric mandrel (Figure 6-37) is used when an eccentricity is desired on the outside diameter of a workpiece.

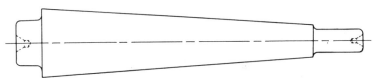

Figure 6-34 Straight lathe mandrel. *(Courtesy Cincinnati Milacron Co.)*

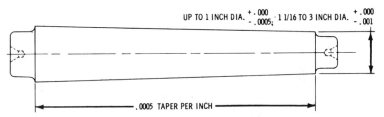

Figure 6-35 Tapered mandrel. *(Courtesy Cincinnati Milacron Co.)*

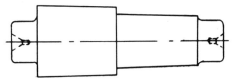

Figure 6-36 Double-diameter mandrel. *(Courtesy Cincinnati Milacron Co.)*

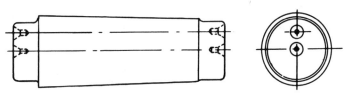

Figure 6-37 Eccentric mandrel. *(Courtesy Cincinnati Milacron Co.)*

Summary

The screw-cutting lathe is the oldest and most important of all the mechanical tools. Practically all of our modern machine tools have been developed from it. The lathe produces cutting action by rotating the workpiece against the edge of a cutting tool.

Lathe work is usually in the shape of a cylinder because the lathe produces a surface that is curved in one direction and is straight in the other.

The maximum size of work that can be handled by the lathe is used to designate the size of the lathe. This is the diameter and length of the work. The manufacturer usually uses the term *swing* to designate the maximum diameter of the work that can be machined in the lathe.

Lathes vary widely from small bench models to the large gapbed type and special-purpose machines. Tool room lathes are usually used to machine small parts. The large lathes are used in high-production work. Some examples of these are turret lathes, multiple-tool lathes, and multiple-spindle lathes. A lathe is made up of many parts, such as the bed, headstock, tailstock, carriage, feed mechanism, and thread-cutting mechanism.

Review Questions

1. What determines the lathe size to use for a particular job?
2. Name the main components of a lathe.
3. Name three types of lathes generally used in high-production, heavy-duty work.
4. What is the carriage section of a lathe?
5. Name the various types of chucks used on lathes.
6. The engine lathe produces cutting action by _____ the workpiece against the edge of the cutting tool.
7. What does "swing" mean?
8. Engine lathes vary widely from the small bench lathe or tool room lathe to the large _____lathe.
9. What are the ways of a lathe?
10. What is a dog?
11. What is meant by dead center?
12. What is the purpose of the tailstock?
13. What does the lathe carriage do?
14. What is a compound rest?

15. Identify the following:
   A. Tool rest
   B. Apron
   C. Lead Screw
   D. Reversing gears
   E. Quick-change gearbox
   F. Threaded spindle nose
   G. Collet
   H. Face plate

16. What is the difference between the three-jaw and four-jaw chuck?
17. What is another name for the four-jaw chuck?
18. Describe live center.
19. What is a dead center?
20. What is the difference between a female and a ball center?
21. Describe the half center and the crotch center.
22. What is a straight lathe mandrel?
23. Distinguish between:
   A. Straight dog
   B. Clamp dog
24. What is a bent-tail dog used for?
25. How can you identify an eccentric mandrel?

Chapter 7

Lathe Operations

Proficiency in lathe operations involves more than just "turning" metal. The machinist should also have an understanding of necessary maintenance procedures and preliminary operations necessary to keep the lathe in good working condition.

Precision of any lathe, regardless of size, is greatly dependent upon the rigidity of the base under the lathe bed. Lighting should be adequate, and there should be enough space around the area for freedom in movement.

The engine lathe should be set level and remain level. If the lathe is not leveled, it will not set squarely on its legs, and the lathe bed will be twisted. If the headstock and V-ways are out of alignment, the workpiece will be tapered.

Wide metal shims and a sensitive level at least 12 inches in length are preferred for leveling a lathe. The bubble of the level should show a distinct movement when a 0.003-inch shim is placed under one end of the level. Both the headstock end and the tailstock end of the lathe should be leveled across the bed ways. Leveling should be repeated after the legs have been bolted down. Most machinists check the lathe for the level position regularly, and whenever the lathe is expected to be used for a long period for a particular job. The level position should also be checked before heavy work is attempted, and whenever the lathe is moved to a new shop location.

Premachining Operations

Good finished work can be turned out on the lathe if the job is planned in advance. All parts are manufactured in a given operational sequence. Lathe operations are not difficult if the work is planned properly.

Cutting Speeds

Lathes are provided with a range of speeds that is ample to meet all conditions. Success in lathe work depends on the proper cutting speed. A cutting speed that is too slow wastes time and leaves a rough finish; a speed that is too high burns the cutting tool.

The cutting speed (tangential velocity) is the speed in feet per minute at which the surface of the work passes the cutting tool. The spool of thread analogy can be used to represent tangential speed as applied to lathe operations (Figure 7-1). The familiar operation of pulling thread off a spool can be used to represent the speed of the revolving surface of the work in a lathe as it passes the cutting tool.

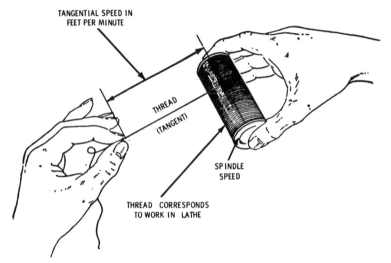

Figure 7-1 The surface speed of the spool as the thread is unwound is comparable to the surface speed of the workpiece as the metal is removed by the cutting tool.

The cutting speed (feet per minute) at which the work passes the cutting edge of the tool is the chief factor in determining selection of the proper headstock spindle speed (revolutions per minute, or rpm) of the lathe for a given material. Proper cutting speed lengthens tool life and results in a better finish. The recommended cutting speeds for the various materials and operations are given in Table 7-1.

Cutting speeds can be calculated by the following formulas:

$$\text{Spindle rpm} = \frac{\text{cutting speed (ft/min.)} \times 12}{\text{circumference of work } (\pi D)}$$

$$= \frac{3.82 \times \text{cutting speed}}{\text{diameter of work}}$$

$$\text{Cutting speed} = \frac{\text{rpm} \times \text{diameter of work}}{3.82}$$

Table 7-1   Cutting Speeds (Feet per Minute) for High-Speed Steel Tools

C.M.M. 320, 1117, 1112, 1018	Cast Iron—60% Steel	S.A.E. 3115, 3310, 4615, 8617	S.A.E. 81-B-45, 8650, 3150, L.G.T.S., H.G.T.S., H.S.S.	S.A.E. 3450, 4340	Bronze	Ampco Bronze	
30	20	20	20	20	60	20	Tap
40	34	36	34	24	20	20	Ream
80	58	62	50	36	198	50	Drill solid
90	58	62	54	38	134	54	Rough bore
90	64	66	66	48	180	66	Finish bore
100	70	66	62	42	180	62	Rough turn (memorize cut speed in italics)
120	78	80	66	48	218	66	Finish turn
100	70	72	66	48	180	66	Rough face
80	78	70	66	48	148	66	Cut off
120	100	108	90	68	218	98	Finish face
30	28	30	28	20	40	28	Knurl (use .010" or .020" feed)
120	80	108	98	68	120	98	Center drilling
40	20	25	20	20	45	20	Chasing or threading

Note: Increase cutting speed approximately three times for carbide tools.
Courtesy Cincinnati Milacron Co.

Feed

Feed is the distance (in thousandths of an inch) that a tool advances into the work during one revolution of the headstock spindle. The maximum feed that is practical should always be used, considering the work, finish and accuracy required, and the rigidity of the work. Maximum efficient production is obtained only by familiarity with the work, the tools, and the lathe. The following formulas can be used to determine feed and time consumed for the operation.

$$F = \frac{L}{N \times T}; \quad \text{and} \quad T = \frac{L}{F \times N}$$

where the following is true:

F is equal to feed in thousandths of an inch.

L is equal to length of cut.

N is equal to rpm.

T is equal to time in minutes.

Depth of Cut

To avoid lost time and wasted horsepower, the maximum amount of metal should be removed per cut per horsepower of the lathe. Cutting speed, feed, and depth of cut determine the amount of metal that is removed. As the operator becomes more familiar with the lathe and the work, the operator is more able to judge and use the maximum depth of cut.

Checking Alignment of Lathe Centers

If the lathe centers are not properly aligned, a tapered surface rather than a cylindrical surface will result. The lathe centers can be checked for alignment prior to turning by placing the dead center of the tailstock in close contact with the live center (Figure 7-2). If the tailstock center does not line up, loosen the tailstock clamp bolt, and move the tailstock top in the proper direction by adjusting the tailstock setover screws to an approximate setting.

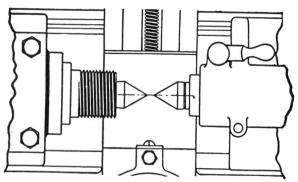

Figure 7-2   Preliminary method of checking alignment of lathe centers.

The final test for alignment of centers is illustrated in Figure 7-3. Turn two 1.5-inch collars (*A* and *B*) on a piece of round stock that has been centered. After a fine finishing cut, measure collar *A*. Then, without changing the caliper setting, measure collar *B*. Compare the diameters of the two collars. If the diameters are not the same, the centers are not in alignment, and the tailstock must be adjusted in the proper direction.

Tailstock adjustment is illustrated in Figure 7-4. The top of the tailstock can be set over by releasing one of the tailstock adjustment screws and tightening the opposite adjustment screw. The marks on the end of the tailstock top and bottom indicate the relative positions of the top and bottom (see Figure 7-4). The alignment of these marks cannot be depended upon for fine, accurate work. The alignment test should be made to make certain that the lathe centers are aligned properly.

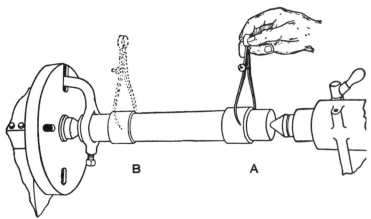

B   **A**

Figure 7-3   Final test for alignment of lathe centers for straight turning.

Centering the Workpiece

Center drilling is a very important premachining operation because the piece to be turned is supported on the lathe centers. The center holes are prepared by drilling and countersinking a hole in each end of the workpiece. The countersunk holes should have the same angle as the lathe centers that fit into them. The standard angle is 60°, and the center hole should be deep enough that the point of the lathe center does not strike the bottom of the hole. A combination drill and countersink is used for this purpose. The center drill should

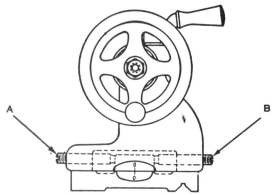

Figure 7-4 Adjustment of the tailstock for alignment of the lathe centers by adjusting the setscrews (A and B).

penetrate into the work ¼₄ (0.15625) inch past the 120° counter-sink (Figure 7-5) for the lathe center to fit properly (Figure 7-6).

In the center-drilling operation, the stock is usually placed in the universal chuck. Not more than four or five times the diameter of the stock should be permitted to extend from the chuck. If the work is longer, a steady rest can be used for support—relieving the chuck jaws of excessive stress. The combination drill and countersink is held in a drill chuck mounted in the tailstock spindle (Figure 7-7).

Basic Operations

The first and simplest exercise for the beginner in lathe operations is turning a piece of round stock to a given diameter. The work is mounted between lathe centers, and a lathe dog is used to drive the work.

Facing

After mounting the work between centers, the facing operation is performed. If an end is to be faced, the operation is more easily per-formed with the work mounted in the universal chuck. The facing tool is used to machine the ends of the work and to machine shoul-ders on the work.

The facing tool should be set with the point exactly on center (Figure 7-8). The cutting edge of the facing tool should be set with the point of the tool slightly toward the work (Figure 7-9).

Roughing cuts with the facing tool can be made either toward or away from the center. However, the finishing cuts must be made by

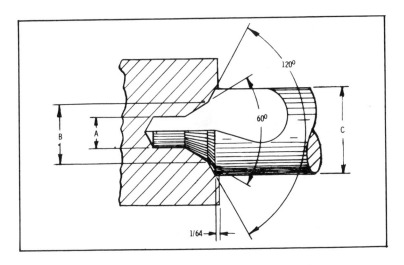

Stock Size	Drill No.	A = Drill Dia.	B = Bell Dia.	C = Body Dia.
$^1/_8 - ^1/_4$	11	$^3/_{64}$	0.100	$^1/_8$
$^1/_4 - ^7/_{16}$	12	$^1/_{16}$	0.150	$^3/_{16}$
$^3/_8 - ^5/_8$	13	$^3/_{32}$	0.200	$^1/_4$
$^1/_2 - 1^1/_4$	14	$^7/_{64}$	0.250	$^5/_{16}$
$1^1/_4 - 1^3/_4$	15	$^5/_{32}$	0.350	$^7/_{16}$
$1^3/_8 - 2$	16	$^3/_{16}$	0.400	$^1/_2$
$1^2/_2 - 2^1/_2$	17	$^7/_{32}$	0.500	$^5/_8$
Over $2^1/_2$	18	$^1/_4$	0.600	$^3/_4$

Figure 7-5 Dimensions of holes that are center drilled with a combination drill and countersink. *(Courtesy Cincinnati Milacron Co.)*

Figure 7-6 Center-drilled holes: (A) properly drilled; (B) incorrectly drilled (too deep); and (C) incorrectly drilled (not deep enough).

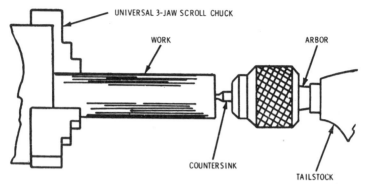

Figure 7-7 Center drilling in round stock.

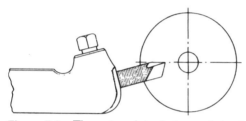

Figure 7-8 The point of the facing tool should be set exactly on center.

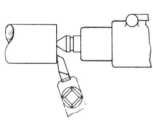

Figure 7-9 The cutting edge of the facing tool should be set with the point turned slightly toward the work.

moving the point of the tool from the center to the outer edge of the cylinder. The point of the facing tool should be rounded slightly on an oilstone to give a smooth finish.

After one end is faced, the work should be reversed and the operation repeated, facing off just enough metal to bring the work to the specified length.

Turning

After the ends have been faced and center-drilled, the work should be turned to the specified diameter.

Mounting on Centers

In straight turning, a bent-tail dog is used to drive the work. It is important to select a lathe dog of correct size for the work and the slot of the driver plate. The tail of the dog should not bind at any point in the slot, nor should it touch the slot, except while driving the work—it should rock back and forth in the slot when mounted properly (Figure 7-10).

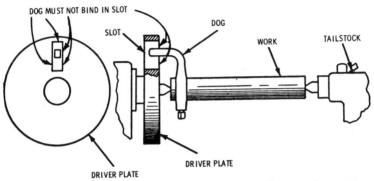

Figure 7-10 Workpiece mounted between the lathe centers with a lathe dog of correct size. The tail of the dog must be free enough to rock back and forth in the slot in the driver plate.

The live center is used only for support of the work, and turns with the work. Therefore, it requires no lubrication. However, the dead center must be lubricated, as the dead center does not turn with the work, and a large quantity of heat can be generated. If the work is mounted too tightly or if lubrication is lacking, the tailstock center will burn. The work can be adjusted for play by means of the tailstock handwheel. The tailstock spindle should be clamped in position at the point where the lathe dog ceases to "chatter" when the lathe has been started.

Height of Cutting Tool

If the height of the cutting edge of a lathe tool coincides with the lathe axis (that is, the lathe centers), the tool is said to be *on center*. If higher or lower than the lathe axis, it is *above* or *below center*. The turning tool is usually set at a slight angle so that the tool will not dig into the work. If it should slip in the tool post, it will move away from the work (Figure 7-11).

Theoretically, the correct height of a cutting tool is at a point where a line through the front edge of the cutting tool coincides with a line tangent to the work. This corresponds to *zero front clearance* (Figure 7-12). The theoretically correct height setting is ideal for a perfectly ground cutting tool. However, as the tool becomes dull, the

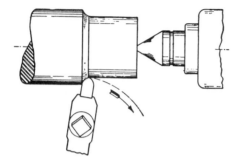

Figure 7-11 Turning tool set at a slight angle so that it will not dig into the work if it should work loose in the tool post.

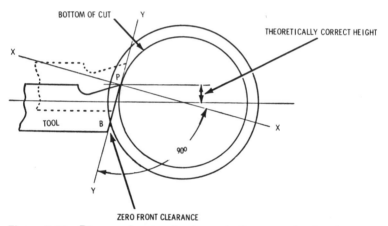

Figure 7-12 Diagram showing the theoretically correct height of a cutting tool. If the front edge of the tool (PB) coincides with the line tangent to the work (YY), the tool is set at the correct theoretical height.

point rounds off slightly. Hence, the correct setting is actually slightly below the theoretically correct height setting.

The theoretically correct height setting decreases as the diameter of the work decreases (Figure 7-13). As shown in Figure 7-13, the tool cannot cut at the smaller diameter because the original theoretically correct height setting for the large diameter is no longer correct because more metal is removed in turning the work.

In practical straight turning between centers, the cutting edge of the tool should be set at about 5° above center (3/64, or 0.046875, inch per inch diameter of the work) as shown in Figure 7-14. Taper turning, boring, and thread-cutting also require the cutting tool to

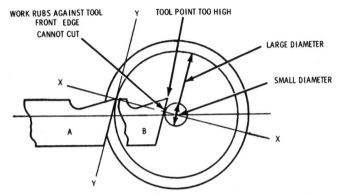

WORK RUBS AGAINST TOOL
FRONT EDGE
CANNOT CUT

Y

TOOL POINT TOO HIGH

LARGE DIAMETER

SMALL DIAMETER

X

A

B

X

Y

Figure 7-13 A decrease in the theoretically correct height setting should be made with a decrease in the diameter of the work. Position A, which is correct for a larger diameter, is incorrect for a smaller diameter because the tool cannot cut at position B.

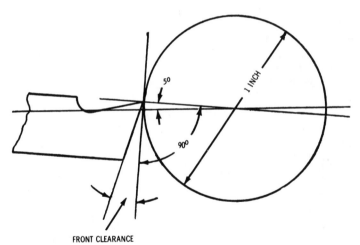

5°

1 INCH

90°

FRONT CLEARANCE

Figure 7-14 Correct height setting of a cutting tool for straight turning.

be set "on center." Materials that require the cutting tool to be set on center for turning are brass, copper, and other tenacious metals.

Roughing Cuts
The roughing tool is used to remove most of the metal. This saves time and leaves only a small amount of metal to be removed by the

finishing cut. Depending on the capacity of the lathe, only one roughing cut may be necessary. Of course, on a long, slender piece of work, the work may spring or break if too deep a cut is taken.

The roughing tool should project no more than necessary from the toolholder. Likewise, the toolholder should project only a short distance from the tool post.

After the roughing tool has been mounted in the tool post, the cross slide should be used to barely "touch up" the work. Then, the cutting tool should be moved to the right-hand side, past the end of the rotating workpiece, and back into the workpiece—using the carriage handwheel. Just after the cutting tool enters the work, engage the power feed for the roughing cut from right to left. Disengage the power feed at about $\frac{1}{8}$ (0.125) inch from the left-hand end of the cut, and feed the tool to the end of the cut (or shoulder) with the carriage handwheel. *Never* feed into a shoulder with the automatic feed because the carriage cannot be stopped at the correct point every time. Thus, both the work and the tool could be ruined. Also, allow ample clearance between the end of the cut and the lathe dog.

After completing the cut with the carriage handwheel, return the carriage to the starting point at the right-hand side. Adjust the cross slide micrometer dial for the depth of the next cut, and repeat the procedure until the workpiece is turned to the desired diameter. Leave approximately $\frac{1}{32}$ (0.0312) inch for finishing the work smoothly to the specified size.

The cutting tool cannot cut to the extreme left-hand end of the work because of the lathe dog. The workpiece must be removed from the lathe, reversed, and placed in the lathe to machine the other end (Figure 7-15). Shim stock should be used under the lathe dog to keep the setscrew from marring the turned end.

Finishing Cuts

After the rough turning operation has been completed, the finishing tool can be mounted in the toolholder (Figure 7-16). Care must be exercised in setting the cutting tool for the finishing cut. If too much metal is removed, the diameter of the work will be too small, and the work will be ruined.

The depth of the finishing cut can be determined from the measurement taken of the rough-turned work. It must be remembered that outside calipers are not precision measuring tools and are reliable only to within approximately 0.005 inch on work that is rough turned and not machined to finished size. Using the hand cross feed, the nose of the finishing tool should be brought in light

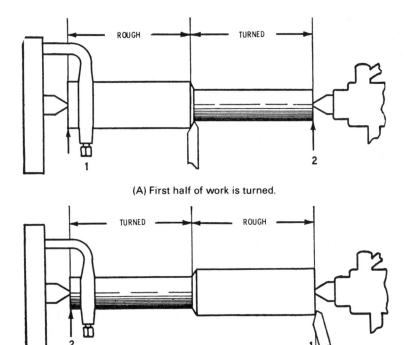

(A) First half of work is turned.

WORK REVERSED

Figure 7-15   Reversing the workpiece to complete the turning operation between centers.

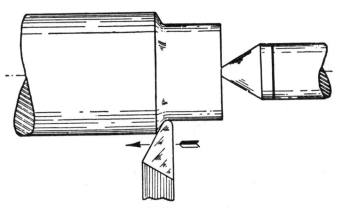

Figure 7-16   Application of a finishing tool.

contact with the rotating workpiece. Note the reading on the cross-feed micrometer dial, and turn the carriage handwheel to the right until the tool is clear of the workpiece. Feed the cutting tool into the work with the hand cross feed—a distance equal to one-half the difference between the rough-turned diameter and the specified finished size.

After selecting the proper cutting speed on the quick-change gearbox, the automatic carriage feed is engaged for the finish cut. After moving a short distance past the end of the work, stop the lathe and measure the diameter with a micrometer caliper. If the work is oversize, return the carriage to the right-hand side and adjust the cross-feed micrometer dial for the correct depth of cut. Proceed with the automatic carriage feed as close to the lathe dog as possible. Stop the lathe, remove the work, place the lathe dog on the opposite end of the workpiece, and complete the work to the specified size.

Several other turning tools can be used for machining work mounted between centers. Whenever possible, the right-hand facing, roughing, and finishing tools should be used because they move from right to left, which avoids placing extra friction or thrust against the dead center of the tailstock. The lathe headstock is mounted more sturdily, and the live center turns with the work. Therefore, these parts are more capable of withstanding the heavy thrusts involved in turning.

In many instances, the left-hand tools (used for machining work from left to right) can be used more advantageously (Figure 7-17). The round-nose tool can be used for turning in either direction. It is a convenient tool for reducing the diameter of a shaft in a central portion of the work (Figure 7-18).

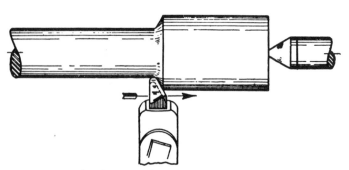

Figure 7-17   Application of a left-hand tool.

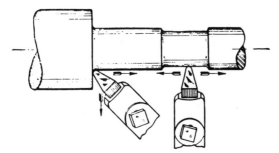

Figure 7-18 Application of a round-nose tool. It is ground flat on top so that it can be used for turning in either direction of feed.

Cutoff Tool

In many instances, stock is cut slightly longer than the finished dimension to allow for center drilling and facing to a finished length. Frequently, several different parts are machined on a single piece of stock and cut apart later. The cutoff tool is used for these purposes.

The cutoff tool should be mounted rigidly, with the carriage clamped to the lathe bed to keep the cutoff tool from moving either to the right or left. The tool should be fed into the work at right angles to the axis of the workpiece. It should be set with the point of the tool exactly on center (Figure 7-19). Cast iron requires no lubrication, but steel requires that the work be flooded with oil.

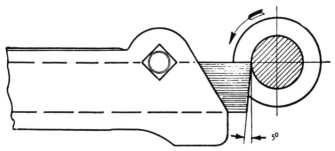

Figure 7-19 Application of the cutoff tool. The tool point should be set exactly on center.

Work Mounted on the Faceplate

An engine lathe should have a large faceplate for mounting work that cannot be held conveniently in a chuck. The work is secured to the faceplate by clamp bolts that are anchored in the faceplate slots.

This type of mounting is illustrated by the disc mounted directly to the faceplate, as shown in Figure 7-20. If the surface of the disc in contact with the faceplate is machined before mounting, a piece of paper should be placed between them to prevent slipping. However, if the rough casting is clamped to the faceplate, shims should be used to make the casting run true with the faceplate (that is, surface *A* should be parallel to surface *B*). Of course, the work should be centered carefully with respect to the hub (see Figure 7-20).

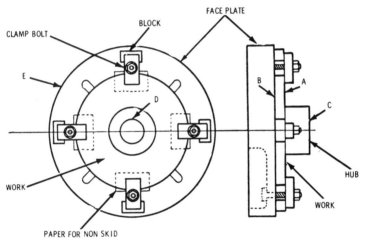

Figure 7-20   Workpiece mounted on the faceplate for turning.

A round-nose tool should be used to face the hub surface C. Then, the hole for the shaft D should be finished to the specific diameter with a boring tool.

If the hole for the shaft is larger than the central hole in the faceplate, parallel blocks should be used to set off the casting from the faceplate, so that the cutting tool will have clearance and will not contact the faceplate when finishing a cut. After this operation has been completed, the work can be removed from the faceplate and mounted on a mandrel, and the machining operation can be continued with the work mounted between centers. The rim should be turned to the finished diameter and the surface *B* faced. If desired, the other surface *A* and the hub can be machined on the mandrel. A "driver" should be mounted between the work and the faceplate to take the thrust and to keep the work from slipping when work having a large diameter and requiring a deep cut is being rotated.

Work Mounted on the Angle Plate

Some irregular pieces cannot be machined by attaching them directly to the faceplate. The angle plate (Figure 7-21) has two working sides machined at a 90° angle. In the setup for machining a flanged elbow (Figure 7-22), the work is bolted or clamped to the angle plate, which is attached to the faceplate. Thus, if one finished flange is attached to the faceplate, the two flanged surfaces will be at 90°.

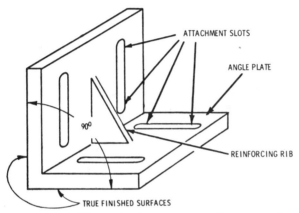

Figure 7-21   The angle plate can be used to mount irregular pieces onto the faceplate for turning.

In actually performing the setup (see Figure 7-22), the angle plate should be mounted on the faceplate so that the distance of the working surface of the angle plate from the lathe center is equal to the distance of one flange face from the axis of the other flange. This distance can be taken from the blueprint. If the elbow is centered laterally by adjusting its position on the angle plate, the facing cut will be concentric. It may be desirable to attach a counterweight, especially for light work, so that spindle speed can be increased.

Work Mounted on the Carriage

Workpieces too large to mount on a faceplate, or in a chuck, can be anchored to the lathe carriage, and the cutting tool mounted between the lathe centers (Figure 7-23). Large pieces (a pump cylinder, for example) can be bored by this method. After bolting the cylinder A to the saddle B of the lathe and mounting the cutter bar C between lathe centers, the cylinder can be bored by engaging the power feed that moves the cylinder toward the headstock.

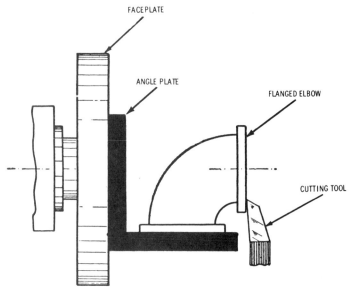

Figure 7-22   Method of mounting an irregular workpiece, such as a flanged elbow, on an angle plate.

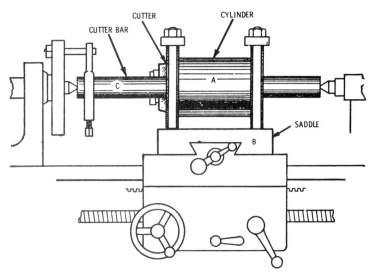

Figure 7-23   Cylindrical workpiece mounted on the lathe carriage for boring.

In some instances, blocks or special fixtures bolted to the carriage can be used to hold the part to be bored. A cylindrical casting (a bushing, for example) can be mounted between wooden blocks bolted across the carriage. The blocks must be cut away to form a circular seat of correct radius, so that the work will be concentric with the boring bar.

The work to be bored should be mounted carefully and accurately on the lathe carriage. Circles should be laid off on each end of the cylinder to show the exact size and position of the bore when completed. The cylinder should be mounted temporarily on the carriage and the boring bar placed between the centers and through the cylinder (Figure 7-24). With a pointed scriber wedged in the cutter slot on the boring bar, the boring bar should be rotated by hand and the cylinder adjusted in position until the scriber point is in contact with the layout circle on the cylinder at all points. This procedure must be repeated on the other end of the cylinder and the hold-down bolts must be tightened to fasten the work onto the carriage securely.

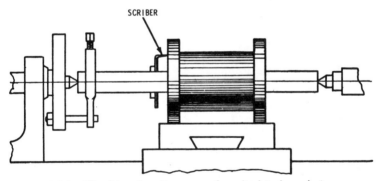

SCRIBER

Figure 7-24 Checking the alignment of a cylindrical workpiece mounted on the lathe carriage.

Drilling

The drilling operation can be performed very precisely on the lathe. The drill can be held stationary and the work rotated, or the work can be held stationary and the drill rotated. The first method is more accurate than the second. The work is held in the chuck mounted on the headstock spindle, and the drill is held stationary in a drill chuck mounted in the tapered hole of the tailstock spindle. Then, the drill is fed into the work with the tailstock handwheel.

As both the tailstock spindle and the headstock spindle have the same inside taper, the drill chuck arbor can be mounted in the headstock spindle for the second method of drilling (that is, the drill is rotated and the work is held stationary). The crotch center and the drill pad are two convenient accessories that can be mounted in the tailstock spindle for holding the work (Figure 7-25).

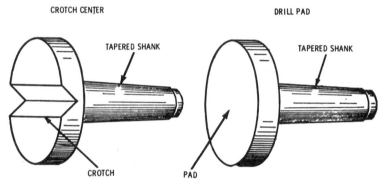

CROTCH CENTER   DRILL PAD

TAPERED SHANK   TAPERED SHANK

CROTCH   PAD

Figure 7-25   Crotch center (left) and drill pad (right).

The crotch center automatically centers round work for cross drilling. The drill pad serves as a table for flat or square work. It is especially valuable for drilling large holes. The work should be held in the left hand and advanced against the drill by turning the tailstock handwheel with the right hand.

Reaming

The lathe can be used to ream holes to a very precise finished diameter. The holder for the reamer is held in the tailstock spindle. The reamer should be positioned at the end of the hole to be reamed. Then, the headstock spindle should be started, and the reamer fed into the hole to the proper depth by means of the tailstock handwheel. Upon reaching the required depth, stop the lathe, and withdraw the reamer from the finished hole.

Holes are usually drilled a few thousandths under size for reaming to the finished size. A small drill of the finished size can be used up to $\frac{1}{16}$ (0.0625) inch in diameter. Drills smaller than the finished size should be used for holes larger than $\frac{1}{16}$ (0.0625) inch in diameter. For drilling and reaming holes, use a center drill, a starting drill, a twist drill $\frac{1}{32}$ (0.03125) inch under the finished size, a twist drill $\frac{1}{64}$ (0.015625) inch under the finished size, and a reamer of the finished size.

For drilling, boring, and reaming holes up to 2⅛ (2.125) inches in diameter, rough-drill the hole approximately ¹⁄₁₆ (0.0625) inch smaller than finished size, and ream to the finished size. For holes larger than 2⅛ (2.125) inches, the work is usually bored to the finished size.

Boring

The removal of stock from a hole in the workpiece while it is being rotated by a lathe is called *boring*. This can be done accurately and efficiently with the proper boring tool, which must be mounted with the point exactly on center. The operator should make certain that there is ample clearance for the boring tool to enter the hole.

The rotating workpiece should be "touched up" by using the cross-feed hand lever, and the tool positioned for a light "cleanup" cut. Then, engage the longitudinal feed and complete the cut. The procedure must be repeated until the correct finished size is reached.

Knurling

The surface of the workpiece can be checked or roughened by rolling depressions into it by means of a knurling tool (Figure 7-26). In the knurling operation, the lathe dog is used to rotate the work. The knurling tool makes indentations in the surface in the form of straight lines or a diamond pattern (Figure 7-27). Knurling provides a better hand grip on the metal and enriches the surface of the workpiece.

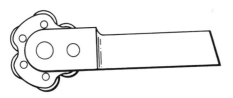

Figure 7-26   A knurling tool.

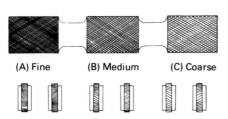

(A) Fine   (B) Medium   (C) Coarse

Figure 7-27   Knurled patterns (above) and knurling wheels (below) showing fine, medium, and coarse knurls. *(Courtesy Cincinnati Milacron Co.)*

The workpiece should be turned to the specified diameter before beginning the knurling operation. Then, the section to be knurled should be turned to the diameter calculated as follows:

Number of teeth on circumference of work = 66.8 × (outside diameter − 0.017)

and

Turned diameter = 0.01496 × number of teeth on circumference of workpiece

The knurling tool should be mounted in the tool post perpendicular to and on the centerline of the workpiece. Feed the cross slide inward until the knurling tool contacts the section to be knurled. The longitudinal feed should be set at 0.010 inch or 0.020 inch on the quick-change gearbox. Then, jog the lathe "on" and "off," while feeding the cross slide in, until the correct depth is attained. The appearance of the knurled surface determines the proper depth. Insufficient depth leaves flats on the diamond pattern; knurling too deeply causes nicks in the surface.

Several precautions should be observed in the knurling operation:

- The axles of the knurling wheels should be kept well lubricated to keep them turning freely.
- In setting the knurling tool, care should be taken to make certain that both wheels contact the surface evenly.
- The knurling tool should be allowed to travel beyond the edge of the workpiece.
- Workpieces having a small diameter should be supported with a steady rest to prevent springing.

Turning an Eccentric

An *eccentric*, by definition, is a circle or sphere not having the same center as another circle partly within or around it. A disc set off-center on a shaft and revolving inside a strap attached to one end of a rod converts the circular motions of the shaft into a back-and-forth motion of the rod. Thus, an eccentric can be used to obtain a reciprocating motion from a rotary motion. The distance between the center of the shaft and the center of the eccentric is the *eccentricity*. This is incorrectly called the *throw*, because the throw is twice the eccentricity.

In machining an eccentric with a hub, the centers of the eccentric should be laid out, the eccentric mounted on a faceplate and centered with respect to the center of the shaft through the hub, and a hole bored for the shaft. Then, the centers of the eccentric should be laid out on a mandrel of correct size for the hole in the eccentric (Figure 7-28). The mandrel should be placed on V-blocks set on a surface gauge and a line *AB* scribed through the mandrel center *O*. This is repeated on the other end of the mandrel. With the dividers set at the same length as the eccentricity, scribe an arc intersecting line *AB* and *C*, which is the center of the eccentric. Repeat the procedure at the other end of the mandrel. Then, the centers for the eccentric *C* can be drilled and countersunk.

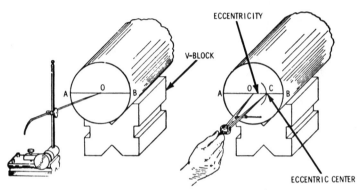

Figure 7-28   A method of locating the centers of an eccentric on a mandrel.

After mounting the eccentric on the mandrel, place the mandrel in the lathe, with the lathe centers engaging the mandrel centers *O*, and machine the hub *D* and both sides *E*. Remove the mandrel and reset in the lathe, with the centers of the eccentric *C* engaging the lathe centers. Machine the face *F* (Figure 7-29).

Taper Turning

A *taper* is a cone-shaped surface—the diameter diminishing uniformly as the distance increases from the largest portion. Taper is measured as *taper (inches) per foot*. Thus, a tapered piece of work that has a 2-inch diameter at one end and a 1-inch diameter at the opposite end is said to have a taper equal to 1 inch per foot of length, or a *1-inch taper*.

A taper can be turned on a piece of work by means of one of the following methods:

- Offset tailstock, thereby setting the lathe centers out of alignment.

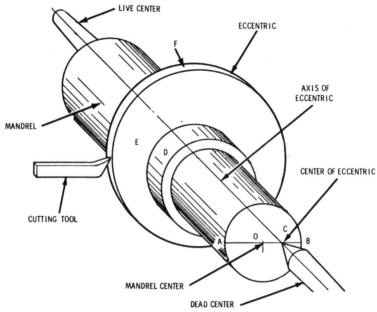

Figure 7-29   Setup for machining an eccentric.

- Compound rest set at an angle.
- Taper attachment, which duplicates a set taper on workpiece.

Theoretically, the offset tailstock method is incorrect, but it can be used within limits (Figure 7-30). The compound rest can be used to turn short tapers and bevels. The taper attachment discussed in Chapter 9 is the approved method for turning work mounted between lathe centers.

Taper per Foot
Length of the work and the difference in diameters of the two ends are the important factors in calculating taper. The formula for determining taper per foot is:

Taper per foot (inches) =

$$\frac{\text{major diameter} - \text{minor diameter} \times 12}{\text{length (inches)}}$$

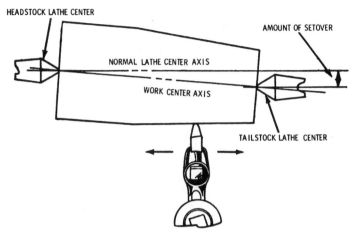

HEADSTOCK LATHE CENTER

AMOUNT OF SETOVER

NORMAL LATHE CENTER AXIS

WORK CENTER AXIS

TAILSTOCK LATHE CENTER

Figure 7-30   Details of taper turning by the tailstock setover method. Note the direction of setover and direction of the taper.

For example, a piece of stock 10 inches in length is 3 inches in diameter at the large end and 2½ inches in diameter at the small end. Calculate the taper per foot by using the following formula:

$$\text{Taper per foot} = \frac{3 - 2.5}{10} \times 12$$

$$= \frac{0.5}{10} \times 12 = 0.6 \text{ inch}$$

Calculating Tailstock Setover

For a given taper, the tailstock must be set over a distance equal to one-half the total amount of taper for the entire length of the workpiece. If the tailstock setover is unchanged for workpieces of different lengths, the pieces will not have the same taper (Figure 7-31).

The distance that the top of the tailstock must be offset depends on the desired taper per foot and the overall length of the workpiece. On a typical lathe, the upper part of the tailstock is locked in position by two headless setscrews (A and B in Figure 7-32) located on the front and back of the base casting. To offset the tailstock, loosen the tailstock clamping nut, loosen the setscrew on the side toward which the tailstock is to be moved, and tighten the opposite setscrew. The index line C at the handwheel end of the tailstock can

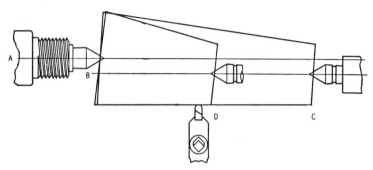

Figure 7-31 Note the effect of the length of stock on taper per foot when the offset tailstock method is used.

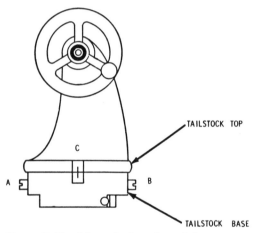

Figure 7-32 Note the location of the setscrews for offsetting the tailstock to turn a taper.

be used to indicate the amount of setover. The amount of tailstock setover can also be measured with a steel rule at the lathe centers (Figure 7-33).

The following formulas can be used to calculate the amount of tailstock setover.

When the taper per foot is given:

$$\text{Setover} = \frac{\text{taper per foot}}{2} \times \frac{\text{length of taper in inches}}{12}$$

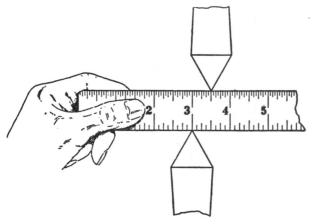

Figure 7-33 A method of measuring the amount of tailstock offset.

When the entire length of the piece is to be tapered, and the diameters at the ends of the tapers are given:

$$\text{Setover} = \frac{\text{major diameter} - \text{minor diameter}}{2}$$

When only a portion of the piece is to be tapered, and the diameters at the ends of the tapered portion are given (see Note):

$$\text{Setover} = \frac{\text{total length of work}}{\text{length to be tapered}}$$

$$\times \frac{\text{major diameter} - \text{minor diameter}}{2}$$

Note
A taper greater than 1 inch per foot on stock 6 inches or less in length is usually turned by tapering only a portion of a longer piece of stock, removing the waste portion after completing a taper. This avoids a small inaccuracy that results from misalignment of the angle of the lathe centers and the angle of the countersunk holes in each end of the stock.

Taper Turning with the Compound Rest
This method is usually used for turning and boring short tapers and bevels—especially for bevel gear blanks and for die and

pattern work. The compound rest should be set at the proper angle and the taper machined by turning the compound-rest feed screw by hand (Figure 7-34). All angular or bevel turning should be checked with a gauge as the operation progresses, because it is difficult to set the compound-rest swivel accurately for the desired taper.

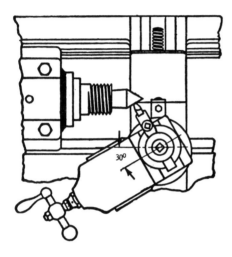

Figure 7-34 The compound rest method of machining a 60° lathe center.

Filing on the Lathe

You may have to resort to filing to produce a good finish or to bring the work down to size. However, filing does not replace finishing by turning. If the work is within one or two thousandths of finished size, and if specifications demand that the work be closer to finished size, the work can be filed while in the lathe.

When it is necessary to file on the lathe, the file should be held with the handle in the palm of one hand with the thumb on the top. The thumb of the other hand should be placed on top near the end of the file with the fingers curled over the end and gripping the file from underneath (Figure 7-35).

Only light pressure should be applied on the forward stroke, releasing the pressure on the back stroke without lifting the file from the work and taking slow, even strokes. The file should not be held in one place on the work, but should be moved diagonally across the work, alternating the stroke from left to right and right to left.

Figure 7-35 Filing a workpiece mounted between centers on the lathe. *(Courtesy Nicolson File Company)*

Summary

The precision of any lathe depends on the rigidity of the base under the lathe bed. There should be enough space around the area for freedom of movement and adequate lighting to perform all operations without eyestrain.

Good finished work can be turned out on the lathe if the job is planned in advance. All parts are manufactured in a given operational sequence. Lathe operations are not difficult if the work is planned properly.

Success in the work depends on the proper cutting speed. A cutting speed that is too slow can waste time and material, and a speed that is too high can burn the cutting tools. Lathes are provided with a range of speeds that is ample to meet all conditions. The cutting speed is the speed in feet per minute at which the surface of the work passes the cutting tool. The cutting speed at which the work passes the cutting edge of the tool is the chief factor in selecting the proper headstock spindle speed for a given material.

Center drilling is a very important premachining operation because the piece to be turned is supported on the lathe centers. The center holes are prepared by drilling and countersinking a hole in each end of the workpiece.

Center holes are very important because the piece to be turned is supported on the lathe center with the help of these holes. The standard angle is 60°, and the center hole should be deep enough so that the point of the lathe center does not strike the bottom of the hole. After the ends have been faced and center drilled, the work should be turned to the specific diameter.

Theoretically, the correct height of a cutting tool is at a point where a line through the front edge of the cutting tool coincides with a line tangent to the work. This corresponds to zero front clearance. The theoretically correct height setting is ideal for a perfectly ground cutting tool. However, as the tool becomes dull, the point rounds off slightly. Hence, the correct setting is actually slightly below the theoretically correct height setting.

In practical straight turning between centers, the cutting edge of the tool should be set at about 5° above center, or $\frac{3}{64}$ (0.046875) inch per inch diameter of the work. Taper turning, boring, and thread-cutting also require the cutting tool to be set "on center."

An engine lathe should have a large faceplate for mounting work that cannot be held conventionally in a chuck. Some irregular pieces cannot be machined by attaching them directly to the faceplate. Workpieces too large to mount on a faceplate (or in a chuck) can be anchored to the lathe carriage, and the cutting tool mounted between the lathe centers.

The removal of stock from a hole in the workpiece while it is being rotated by a lathe is called boring. The surface of a workpiece can be checked or roughened by rolling depressions into it by means of a knurling tool. In machining an eccentric with a hub, the centers of the eccentric should be laid out, the eccentric mounted on a faceplate and centered with respect to the center of the shaft through the hub, and a hole bored for the shaft. A taper is a cone-shaped surface—the diameter diminishing uniformly as the distance increases from the largest portion. Taper is measured as taper (inches) per foot.

Review Questions

1. Why are center holes drilled in pieces to be turned?
2. Why is the proper speed important in working with a lathe?
3. What is the standard angle for center holes?
4. What happens if the lathe centers are not properly aligned?
5. Why should a lathe be level?
6. What is meant by "feed"?
7. Why is the center-drilled hole important on a piece of stock?
8. What is a facing operation on a lathe?
9. Why is the turning tool set at a slight angle?
10. What is the next step after completing the cut with the carriage handwheel?
11. How does a cutoff tool differ from a turning tool?
12. How is work held on a faceplate for turning?
13. What is the purpose of an angle plate?
14. How can drilling be done on a lathe?
15. How is reaming done on a lathe?
16. What is eccentricity?
17. How do you calculate tailstock setover?
18. How do you turn an eccentric?
19. What is a knurling tool?
20. What does the term "taper per foot" mean?

Chapter 8

Cutting Screw Threads on the Lathe

The screw thread is a very important mechanical device used for adjusting, fastening, and transmitting motion. Thread-cutting in the engine lathe requires a thorough knowledge of thread-cutting principles and procedures.

Screw Thread Systems

The threading operation actually involves cutting a helical groove of definite shape or angle, with a uniform advancement for each revolution, either on the surface of a round piece of material, or inside a cylindrical hole. Threads are either right-hand threads (advanced clockwise) or left-hand threads (advanced counterclockwise).

A number of thread systems are in use, which can result in confusion for a mechanic or machinist. Therefore, a mechanic or machinist should be familiar with the standard systems and their adaptations.

Screw Thread Forms

The most-used forms of screw threads have symmetrical sides inclined at equal angles with a vertical centerline through the apex of the thread. The *Unified, Whitworth*, and *Acme* forms are examples. The *Sharp-V* was an early form of thread that is now used only occasionally. The symmetrical threads are easy to manufacture and are widely used as general-purpose fasteners.

In addition, the so-called *translation threads* are used to move, or translate, machine parts against heavy loads. Thus, a stronger thread is required. Some of these translation threads that are widely used are the *square, Acme*, and *buttress* threads. The square thread is the most efficient of these threads. However, it is the most difficult to cut because of its parallel sides, and it cannot be adjusted to compensate for wear. The Acme thread is only slightly weaker and less efficient, and it has none of the disadvantages of the square thread. The buttress thread is used to translate loads in one direction only. It combines the ease of cutting and adjustment of the Acme thread with the efficiency and strength of the square thread.

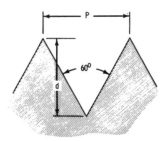

Figure 8-1 Sharp-V thread form. *(Courtesy Machinery's Handbook, The Industrial Press)*

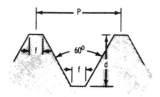

Figure 8-2 American National thread form. *(Courtesy Machinery's Handbook, The Industrial Press)*

Sharp-V Thread

The top and bottom of the thread are theoretically sharp, but they are made slightly flat in actual practice. The sides form an angle of 60° with each other (Figure 8-1).

American National Screw Thread Forms

This form was the standard thread used in the United States for many years (Figure 8-2). The thread form is practically the same as the American Standard for Unified Threads now in use, except for certain modifications.

American Standard for Unified Screw Threads

In 1949 the American Standard Association approved the revision of the *American Standard Screw Threads for Bolts, Nuts, Machine Screws, and Threaded Parts* to cover the Unified screw threads on which the United States, Canada, and Great Britain reached agreement on nominal dimensions and size limits to obtain screw thread interchangeability among the three nations. The Unified threads are now the basic American standard for the fastening types of screw threads (Figure 8-3). These threads have practically the same thread form and are mechanically interchangeable with American National threads of the same diameter and pitch. The principal differences in the two systems are the difference in amount of pitch-diameter tolerance on external and internal threads and the variation of tolerances with size.

In the American National system only the Class 1 external thread has an allowance, but in the Unified system an allowance is provided on both Classes 1A and 2A external threads. Also, the pitch-diameter tolerance of external and internal threads is equal in the American National system, whereas in the Unified system the pitch-diameter tolerance of an internal thread is 30 percent greater than that of the external thread.

In the Unified system, the crest of the external thread can be either flat (preferred by American industry) or rounded (preferred

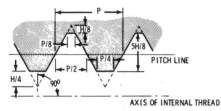

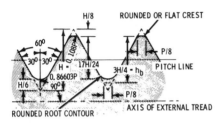

(A) Internal threads.

(B) External threads.

Figure 8-3   Basic form of Unified screw threads.

by British industry). The crest is flat on internal threads. Otherwise, the Unified thread is practically the American National form long used in the United States.

Square Thread
The Acme thread has largely replaced the square thread. Even though the square thread is stronger and more efficient, it is difficult to produce economically because of its perpendicular sides (Figure 8-4).

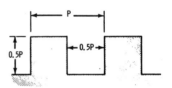

Figure 8-4   Square thread form. *(Courtesy Machinery's Handbook, The Industrial Press)*

American Standard Acme Screw Threads
This standard provides three classes of *general-purpose* Acme threads for use in assemblies that require a rigidly fixed internal thread and have a lateral movement of the external thread limited by its bearing or bearings (Figure 8-5). The standard also provides five classes of *centralizing* Acme threads.

American Standard Stub Acme Screw Threads
A Stub Acme screw thread is used for unusual applications where a coarse-pitch thread with a shallow depth is required.

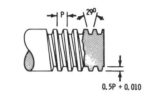

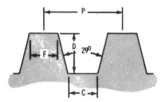

Figure 8-5 Acme thread form.

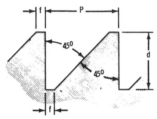

Figure 8-6 Buttress thread form. *(Courtesy Machiner's Handbook, The Industrial Press)*

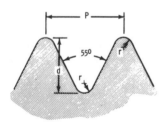

Figure 8-7 Whitworth thread form. *(Courtesy Machinery's Handbook, The Industrial Press)*

American Standard Buttress Screw Threads

This form of thread is used where high stresses along the thread axis in one direction only are involved (Figure 8-6). It is particularly applicable where tubular members are screwed together.

Dardelet Thread

This is a patented self-locking thread designed to resist vibrations and to remain tight. The nut turns freely until it is seated tightly against a resisting surface. Then, it shifts to the locking position because of a wedging action between the tapered crest of the thread of the nut and the tapered root of the thread of the bolt.

Whitworth Standard Screw Thread Form

Since the Unified thread has been standardized, the British Whitworth form of parallel screw thread is expected to be used only for spare or replacement parts (Figure 8-7).

International Metric Thread System

This thread form was adopted at the International Congress for standardization of screw threads at Zurich, Switzerland, in 1898. It is similar to the American Standard thread, except for the depth, which is greater.

American Standard for Unified Miniature Screw Threads

These threads are intended for use as general-purpose fastening screws in watches, instruments, and miniature assemblies or mechanisms. Except for its basic height and depth of engagement, the threads are identical to the American and Unified basic thread form.

Screw Thread Series

Diameter-pitch combinations are grouped to form thread series, which are distinguished by the number of threads per inch as applied to a specific diameter. There are 11 standard series combinations.

Coarse-Thread Series (UNC)

The most commonly-used series in the bulk production of bolts, screws, nuts, and other general applications is the coarse-thread series. It is used for threading materials having a low tensile strength, such as cast iron, mild steel, bronze, brass, aluminum, and plastics (see Table 8-1).

Table 8-1   Coarse-Thread Series, UNC and NC—Basic Dimensions (UNRC)

Sizes	Basic Major Diameter, in.	Threads per Inch	Basic Pitch Diameter, in.	Minor Diameter Exterior Threads, in.	Minor Diameter Interior Threads, in.	Lead Angle at Basic Pitch Diameter Deg	Lead Angle at Basic Pitch Diameter Min.	Area of Minor Diameter, in.²	Tensile Stress Area, in.²
1 (.073)*	0.0730	64	0.0629	0.0538	0.0561	4	31	0.00218	0.00263
2 (.086)	0.0860	56	0.0744	0.0641	0.0667	4	22	0.00310	0.00370
3 (.099)*	0.0990	48	0.0855	0.0734	0.0764	4	26	0.00406	0.00487
4 (.112)	0.1120	40	0.0958	0.0813	0.0849	4	45	0.00496	0.00604
5 (.125)	0.1250	40	0.1088	0.0943	0.0979	4	11	0.00672	0.00796
6 (.138)	0.1380	32	0.1177	0.0997	0.1042	4	50	0.00745	0.00909
8 (.164)	0.1640	32	0.1437	0.1257	0.1302	3	58	0.01196	0.0140
10 (.190)	0.1900	24	0.1629	0.1389	0.1449	4	39	0.01450	0.0175
12 (.216)*	0.2160	24	0.1889	0.1649	0.1709	4	1	0.0206	0.0242
¼	0.2500	20	0.2175	0.1887	0.1959	4	11	0.0269	0.0318
⁵⁄₁₆	0.3125	18	0.2764	0.2443	0.2524	3	40	0.0454	0.0524
⅜	0.3750	16	0.3344	0.2983	0.3073	3	24	0.0678	0.0775
⁷⁄₁₆	0.4375	14	0.3911	0.3499	0.3602	3	20	0.0933	0.1063
½	0.5000	13	1.4500	0.4056	0.4167	3	7	0.1257	0.1419
⁹⁄₁₆	0.5625	12	0.5084	0.4603	0.4723	2	59	0.162	0.182
⅝	0.6250	11	0.5660	0.5135	0.5266	2	56	0.202	0.226
¾	0.7500	10	0.6850	0.6273	0.6417	2	40	0.302	0.334
⅞	0.8750	9	0.8028	0.7387	0.7547	2	31	0.419	0.462

(continued)

Table 8-1   (continued)

| Sizes | Basic Major Diameter, in. | Threads per Inch | Basic Pitch Diameter, in. | Minor Diameter | | Lead Angle at Basic Pitch Diameter | | Area of Minor Diameter, in.² | Tensile Stress Area, in.² |
				Exterior Threads, in.	Interior Threads, in.	Deg	Min.		
1	1.0000	8	0.9188	0.8466	0.8647	2	29	0.551	0.606
1⅛	1.1250	7	1.0322	0.9497	0.9704	2	31	0.693	0.763
1¼	1.2500	7	1.1572	1.0747	1.0954	2	15	0.890	0.969
1⅜	1.3750	6	1.2667	1.1705	1.1946	2	24	1.054	1.155
1½	1.5000	6	1.3917	1.2955	1.3196	2	11	1.294	1.405
1¾	1.7500	5	1.6201	1.5046	1.5335	2	15	1.74	1.90
2	2.0000	4½	1.8557	1.7274	1.7594	2	11	2.30	2.50
2¼	2.2500	4½	2.1057	1.9774	2.0094	1	55	3.02	3.25
2½	2.5000	4	2.3376	2.1933	2.2294	1	57	3.72	4.00
2¾	2.7500	4	2.5876	2.4433	2.4794	1	46	4.62	4.93
3	3.0000	4	2.8376	2.6933	2.7294	1	36	5.62	5.97
3¼	3.2500	4	3.0876	2.9433	2.9794	1	29	6.72	7.10
3½	3.5000	4	3.3376	3.1933	3.2294	1	22	7.92	8.33
3¾	3.7500	4	3.5876	3.4433	3.4794	1	16	9.21	9.66
4	4.0000	4	3.8376	3.6933	3.7294	1	11	10.61	11.08

Courtesy Machinery's Handbook, The Industrial Press

Fine-Thread Series (UNF)

This series is suitable where the coarse series is not applicable for the production of bolts, screws, and nuts. The fine-thread series is used where the length of thread engagement is short and the wall thickness demands a fine pitch (see Table 8-2).

Extra-Fine Thread Series (UNEF)

This series can be used in most conditions applicable for fine threads. It is used where even finer threads are desirable, as for thin-walled tubes, couplings, nuts, or ferrules (see Table 8-3).

Screw Thread Terms

The mechanic or machinist should become familiar with the following terms commonly used in connection with screw thread systems and thread-cutting operations:

- *Allowance*—A prescribed difference between the dimensions of mating parts. "Maximum allowance" is the difference

Table 8-2   Fine-Thread Series, UNF, UNRF, and NF—Basic Dimensions

Sizes	Basic Major Diameter, in.	Threads per Inch	Basic Pitch Diameter, in.	Minor Diameter Exterior Threads, in.	Minor Diameter Interior Threads, in.	Lead Angle at Basic Pitch Diameter Deg	Lead Angle at Basic Pitch Diameter Min.	Area of Minor Diameter, in.^{2}	Tensile Stress Area, in.^{2}
0 (.060)	0.0600	80	0.0519	0.0447	0.0465	4	23	0.00151	0.00180
1 (.073)*	0.0730	72	0.0640	0.0560	0.0580	3	57	0.00237	0.00278
2 (.086)	0.0860	64	0.0759	0.0668	0.0691	3	45	0.00339	0.00394
3 (.099)*	0.0990	56	0.0874	0.0771	0.0797	3	43	0.00451	0.00523
4 (.112)	0.1120	48	0.0985	0.0864	0.0894	3	51	0.00566	0.00661
5 (.125)	0.1250	44	0.1102	0.0971	0.1004	3	45	0.00716	0.00830
6 (.138)	0.1380	40	0.1218	0.1073	0.1109	3	44	0.00874	0.01015
8 (.164)	0.1640	36	0.1460	0.1299	0.1339	3	28	0.01285	0.01474
10 (.190)	0.1900	32	0.1697	0.1517	0.1562	3	21	0.0175	0.0200
12 (.216)	0.2160	28	0.1928	0.1722	0.1773	3	22	0.0226	0.0258
¼	0.2500	28	0.2268	0.2062	0.2113	2	52	0.0326	0.0364
⁵⁄₁₆	0.3125	24	0.2854	0.2614	0.2674	2	40	0.5240	0.0580
⅜	0.3750	24	0.3479	0.3239	0.3299	2	11	0.0809	0.0878
⁷⁄₁₆	0.4375	20	0.4050	0.3762	0.3834	2	15	0.1090	0.1187
½	0.5000	20	0.4675	0.4387	0.4459	1	57	0.1486	0.1599
⁹⁄₁₆	0.5625	18	0.5264	0.4943	0.5024	1	55	0.189	0.203
⅝	0.6250	18	0.5889	0.5568	0.5649	1	43	0.240	0.256
¾	0.7500	16	0.7094	0.6733	0.6823	1	36	0.351	0.373
⅞	0.8750	14	0.8286	0.7874	0.7977	1	34	0.480	0.509
1	1.0000	12	0.9459	0.8978	0.9098	1	36	0.625	0.663
1⅛	1.1250	12	1.0709	1.0228	1.0348	1	25	0.812	0.856
1¼	1.2500	12	1.1959	1.1478	1.1598	1	16	1.024	1.073
1⅜	1.3750	12	1.3209	1.2728	1.2848	1	9	1.260	1.315
1½	1.5000	12	1.4459	1.3978	1.4098	1	3	1.521	1.581

Courtesy Machinery's Handbook, The Industrial Press

between a minimum external and a maximum internal part. "Minimum allowance" is the difference between a maximum external and a minimum internal part. A "negative allowance" is the overlap, or interference, between the dimensions of mating parts. Allowance provides for variations in fit.

- Angle of thread—The included angle between the sides of the thread, measured in an axial plane.

Table 8-3   Extra-Fine-Thread Series, UNEF, UNREF, and NEF—Basic Dimensions

Sizes	Basic Major Diameter, in.	Threads per Inch	Basic Pitch Diameter, in.	Minor Diameter Exterior Threads, in.	Minor Diameter Interior Threads, in.	Lead Angle at Basic Pitch Diameter Deg	Lead Angle at Basic Pitch Diameter Min.	Area of Minor Diameter, in.²	Tensile Stress Area, in.²
12(.216)*	0.3160	32	0.1957	0.1777	0.1822	2	55	0.0242	0.0270
¼	0.2500	32	0.2297	0.2117	0.2162	2	29	0.0344	0.0379
⁵⁄₁₆	0.3125	32	0.2922	0.2742	0.2787	1	57	0.0581	0.0625
³⁄₈	0.3750	32	0.3547	0.3367	0.3412	1	36	0.0878	0.0932
⁷⁄₁₆	0.4375	28	0.4143	0.3937	0.3988	1	34	0.1201	0.1274
½	0.5000	28	0.4768	0.4562	0.4613	1	22	0.162	0.170
⁹⁄₁₆	0.5625	24	0.5354	0.5114	0.5174	1	25	0.203	0.214
⁵⁄₈	0.6250	24	0.5979	0.5739	0.5799	1	16	0.256	0.268
¹¹⁄₁₆*	0.6875	24	0.6604	0.6364	0.6424	1	9	0.315	0.329
¾	0.7500	20	0.7175	0.6887	0.6959	1	16	0.369	0.386
¹³⁄₁₆	0.8125	20	0.7800	0.7512	0.7584	1	10	0.439	0.458
⁷⁄₈	0.8750	20	0.8425	0.8137	0.8209	1	5	0.515	0.536
¹⁵⁄₁₆*	0.9375	20	0.9050	0.8762	0.8834	1	0	0.598	0.620
1	1.0000	20	0.9675	0.9387	0.9459	0	57	0.687	0.711
1¹⁄₁₆*	1.0625	18	1.0264	0.9943	1.0024	0	59	0.770	0.799
1⅛	1.1250	18	1.0889	1.0568	1.0649	0	56	0.871	0.901
1³⁄₁₆*	1.1875	18	1.1514	1.1193	1.1274	0	53	0.977	1.099
1¼	1.2500	18	1.2139	1.1818	1.1899	0	50	1.000	1.123
1⁵⁄₁₆*	1.3125	18	1.2764	1.2443	1.2524	0	48	1.208	1.244
1⅜	1.3750	18	1.3389	1.3068	1.3149	0	45	1.333	1.370
1⁷⁄₁₆*	1.4375	18	1.4014	1.3693	1.3774	0	43	1.464	1.503
1½	1.5000	18	1.4639	1.4318	1.4399	0	42	1.60	1.64
1⁹⁄₁₆*	1.5625	18	1.5264	1.4943	1.5024	0	40	1.74	1.79
1⅝	1.6250	18	1.5889	1.5568	1.5649	0	38	1.89	1.94
1¹¹⁄₁₆*	1.6875	18	1.6514	1.6193	1.6274	0	37	2.05	2.10

Courtesy Machinery's Handbook, The Industrial Press

- *Axis of screw*—The longitudinal central line through the screw from which all corresponding parts are equally distant.
- *Base of thread*—The bottom section of a thread; the largest section between two adjacent roots.
- *Basic*—The theoretical or nominal standard size from which all variations are made.

- *Crest*—The top surface joining the two sides of a thread.
- *Crest Clearance*—Defined on a screw form as the space between the crest of a thread and the root of its component.
- *Depth of engagement*—The depth of a thread in contact, of two mating parts—measured radially.
- *Depth of thread*—The depth, in profile, is the distance between the top and the base of the thread (see Table 8-4).
- *External thread*—A thread on the outside of a member, as on a threaded plug.
- *Helix angle*—The angle made by the helix of the thread at the pitch diameter with a plane perpendicular to the axis.
- *Internal thread*—A thread on the inside of a member, as in a thread hole.
- *Lead*—The distance a screw thread advances axially in one turn. On a single screw thread, lead and pitch are identical; on a double screw thread, lead is twice the pitch, and so on.
- *Length of engagement*—The length of contact between two mating parts—measured axially.
- *Limits*—The extreme dimensions prescribed to provide for variations in fit and workmanship.
- *Major diameter*—The largest diameter of the thread of a screw or nut. The term replaces "outside diameter," as applied to the thread of a screw, and "full diameter," as applied to the thread of a nut.
- *Minor diameter*—The smallest diameter of the thread of a screw or nut. The term replaces "core diameter," as applied to the thread of a screw, and "inside diameter," as applied to the thread of a nut.
- *Neutral space*—The space between mating parts that must not be encroached upon.
- *Number of threads*—The number of threads in one inch of length.
- *Pitch*—The distance from a point on a screw thread to a corresponding point on an adjacent thread, measured parallel to the axis.
- *Pitch diameter*—On a straight screw thread, pitch diameter is the diameter of an imaginary cylinder, the surface of which would pass through the threads at such points as to make equal the width of the threads and the width of the spaces cut by the surface of the cylinder.

Table 8-4 Double Depth (DD) of Threads

Threads per Inch N	V Threads D D	Am. Nat. Form D D U.S. Std.	Whitworth Standard D D	Threads per Inch N	V Threads D D	Am. Nat. Form D D U.S. Std.	Whitworth Standard D D
2	0.86650	0.64950	0.64000	28	0.06185	0.04639	0.04571
2¼	0.77022	0.57733	0.56888	30	0.05773	0.04330	0.04266
2⅜	0.72960	0.45694	0.53894	32	0.05412	0.04059	0.04000
2½	0.69320	0.51960	0.51200	34	0.05097	0.03820	0.03764
2⅝	0.66015	0.49485	0.48761	36	0.04811	0.03608	0.03555
2¾	0.63019	0.47236	0.45545	38	0.04560	0.03418	0.03368
2⅞	0.60278	0.45182	0.44521	40	0.04330	0.03247	0.03200
3	0.57733	0.43300	0.42666	42	0.04126	0.03093	0.03047
3¼	0.53323	0.39966	0.39384	44	0.03936	0.02952	0.02909
3½	0.49485	0.37114	0.35571	46	0.03767	0.02823	0.02782
4	0.43300	0.32475	0.32000	48	0.03608	0.02706	0.02666
4½	0.38438	0.28869	0.23444	50	0.03464	0.02598	0.02560
5	0.34660	0.25980	0.25600	52	0.03332	0.02498	0.02461
5½	0.31490	0.23618	0.23272	54	0.03209	0.02405	0.02370
6	0.28866	0.21650	0.21333	56	0.03093	0.02319	0.02285
7	0.24742	0.18557	0.13285	58	0.02987	0.02239	0.02206
8	0.21650	0.16237	0.15000	60	0.02887	0.02165	0.02133
9	0.19244	0.14433	0.14222	62	0.02795	0.02095	0.02064
10	0.17320	0.12990	0.12800	64	0.02706	0.02029	0.02000
11	0.15745	0.11809	0.11636	66	0.02625	0.01968	0.01939
11½	0.15069	0.11295	0.11121	68	0.02548	0.01910	0.01882
12	0.14433	0.10825	0.10666	70	0.02475	0.01855	0.01728
13	0.13323	0.09992	0.09846	72	0.02407	0.01804	0.01782
14	0.12357	0.09278	0.09142	74	0.02341	0.01752	0.01729
15	0.11555	0.08660	0.08533	76	0.02280	0.01714	0.01673
16	0.10825	0.08118	0.08000	78	0.00221	0.01665	0.01641
18	0.09622	0.07216	0.07111	80	0.02166	0.01623	0.01600
20	0.08660	0.06495	0.06400	82	0.02113	0.01584	0.01560
22	0.07872	0.05904	0.58180	84	0.02063	0.01546	0.01523
24	0.07216	0.05412	0.05333	86	0.02015	0.01510	0.01476
26	0.06661	0.04996	0.04923	88	0.01957	0.01476	0.01454
27	0.06418	0.04811	0.04740	90	0.01925	0.01443	0.01422

Note the following:

$$DD = \frac{1.733}{N} \text{ for V thread}$$

$$DD = \frac{1.299}{N} \text{ for Am. Nat. Form, U.S. Std.}$$

$$DD = \frac{1.28}{N} \text{ for Whitworth Standard}$$

- *Root*—The bottom surface joining the sides of adjacent threads.
- *Screw thread*—A ridge of a desired profile generated in the form of a helix either on the inside or on the outside surface of a cylinder or cone.
- *Side of thread*—The surface of the thread that connects the crest with the root.
- *Tolerance*—The amount of variation permitted in the size of a part. It is the difference between the limits or maximum and minimum dimensions of a given part. Tolerance may be expressed as plus, minus, or as both plus and minus. Total tolerance is the sum of a plus and minus tolerance. Net tolerance is the total tolerance reduced by gauge manufacturing tolerances and wear limits of the part.

Change-Gear Calculations

In the absence of an index chart showing gear combinations for various threads, it is necessary to calculate the proper gears to use for cutting threads. On lathes equipped with tumbler reverse gears, it must first be determined whether the lathe is *even-geared* or *odd-geared*.

On an even-geared lathe, the stud gear revolves at the same speed as the spindle gear (that is, the two gears are equal in size). If the stud gear revolves at any other speed, the lathe is an odd-geared lathe.

Change Gears

Change gears are either *simple* or *compound* in form. In simple gearing, an idler gear is used to transmit motion from the stud gear to the lead screw gear (Figure 8-8). In compound gearing, the idler is combined with another gear of different size. As the two gears are keyed together, the assembly no longer functions as an idler, but as a first-stage ratio reduction between the stud gear and the lead screw gear (Figure 8-9). This "compound gear assembly" is necessary when a large ratio between the stud gear and lead screw gear is required to cut extra-fine threads.

Simple Gearing

The ratio between the number of teeth on the stud gear and the lead screw gear must be determined. The ratio depends on the pitch of the lead screw and the number of teeth to be cut. This is expressed as the following formula:

$$\text{Change-gear ratio} = \frac{\text{threads per inch on lead screw}}{\text{number of threads to be cut}}$$

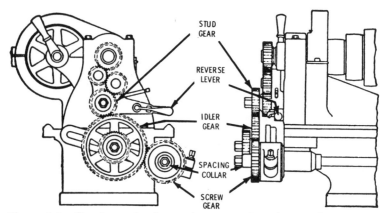

Figure 8-8   Simple gearing for cutting screw threads on a change-gear lathe.

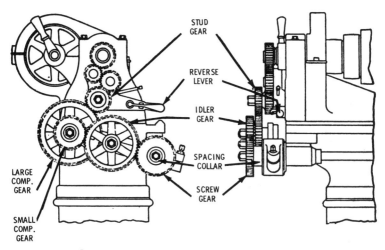

Figure 8-9   Compound gearing for cutting screw threads on a change-gear lathe.

For example, to determine the size of the stud gear and the lead screw gear required to cut 12 threads per inch in a lathe having a lead screw with 8 threads per inch, use the following formula:

$$\text{Gear ratio} = \frac{8}{12}$$

To cut 12 threads per inch, the spindle (or stud gear on an even-geared lathe) must make 12 revolutions to 8 revolutions of the lead screw. Thus, if gears of 12 teeth were available, they would cut the required thread. Because these gears are impossible, multiply each by a common number to obtain the desired gears available in the change-gear set as follows:

1. Stud gear = 8 × 3 = 24 teeth

2. Lead screw gear = 12 × 3 = 36 teeth

Because change gears of various sizes are used, the distance between gears will vary. Therefore, an idler gear must be used to transmit the motion (Figure 8-10). The idler gear does not change the gear ratio.

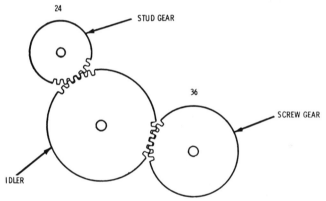

Figure 8-10   Diagram of simple change gearing.

Compound Gearing

When the gears are arranged in a train, they are said to be compounded. As in simple gearing, the gear ratio must be determined between the stud gear and the lead screw gear. To illustrate the necessity for compound gearing, suppose that it is desired to cut 80 threads per inch in a lathe having a lead screw with 8 threads per inch. Use the following formula:

$$\text{Gear ratio} = \frac{8}{80}$$

To cut 80 threads per inch, the spindle (or stud gear in an even-geared lathe) must make 80 revolutions to 8 revolutions of the lead screw. It is impossible to use gears with 8 and 80 teeth to cut the required thread. Multiply the number of teeth by a common number to obtain gears within the range of the change-gear set. Because a 16-tooth gear is usually the lowest number of teeth furnished in the set, multiply by 2 as follows:

1. Stud gear = $8 \times 2 = 16$ teeth

2. Lead screw gear = $80 \times 2 = 160$ teeth

Because a 160-tooth gear is entirely out of range of most equipment, and the required diameter would probably be too large to mesh with the stud gear, compound gearing is necessary so that smaller gears can be used.

For example, to determine the compound gears necessary to cut 80 threads per inch on a lathe having a lead screw with 8 threads per inch, the following procedure can be used:

1. The ratio between the stud gear and the lead screw gear is as follows:

$$\text{Total ratio} = \frac{8}{80}$$

2. Because this is a 10 to 1 ratio, the stud gear must make 10 revolutions to 1 revolution of the lead screw. Factor the total ratio as follows:

$$\frac{8}{80} = \frac{2 \times 4}{8 \times 10}$$

3. Multiply each term by a common number (8, for example), and obtain the required gears as follows:

$$\frac{16}{64} = \frac{32}{80}$$

4. Place these gears in the following order: 16, 64, 32, and 80. Use the 16-tooth gear as the stud gear, use the 64- and 32-tooth gears for the compound gear assembly, and place the 80-tooth gear on the lead screw (Figure 8-11). This gear setup will provide a 10 to 1 ratio or speed reduction between the lathe spindle and the lead screw.

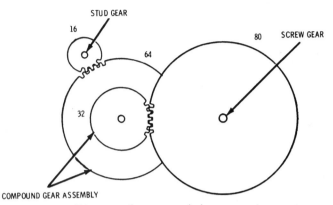

Figure 8-11 Diagram of compound change gearing.

Compound gearing actually consists of two ratios whose product is equivalent to the total ratio as follows:

$$\text{First ratio} = \frac{16}{64} = \frac{1}{4}$$

$$\text{Second ratio} = \frac{32}{80} = \frac{1}{2.5}$$

and

$$\text{Total ratio} = \text{first ratio} \times \text{second ratio}$$

$$= \frac{1}{4} \times \frac{1}{2.5} = \frac{1}{10}$$

When compound gearing is used, the ratio of the compound gears is usually 2 to 1, so that the threads are twice the number per inch as when simple gearing is used. Therefore, the following procedure can be used:

1. Calculate as in simple gearing:

$$\text{Gear ratio} = \frac{8 \text{ (stud gear)}}{40 \text{ (lead screw gear)}}$$

2. Doubling these figures results in 16 teeth for the stud gear and 80 teeth for the lead screw gear, respectively. Selecting, for example, 18- and 36-tooth gears for the compound assembly, the setup would be as follows: A 16-tooth gear for the stud gear, 36- and 18-tooth gears for the compound assembly, and an 80-tooth gear for the lead screw (Figure 8-12). Thus, the following ratios are:

$$\text{Stud gear to lead screw gear ratio} = \frac{16}{80} = \frac{1}{5}$$

$$\text{Compound assembly ratio} = \frac{18}{36} = \frac{1}{2}$$

and

$$\text{Total ratio} = \frac{1}{5} \times \frac{1}{2} = \frac{1}{10}$$

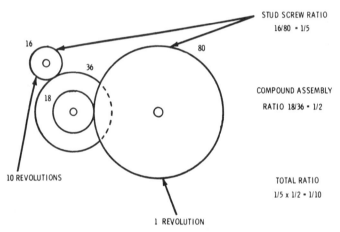

Figure 8-12 Diagram showing use of total ratio in a compound change-gear setup.

Quick-Change Gear Lathes

The quick-change gearbox permits the operator to obtain the various pitches of threads without using loose gears. All lathes equipped with quick-change gearboxes have index plates or charts for setting up the lathe to cut various threads. It is necessary only to arrange the levers on the gearbox for the various threads per inch, as indicated on the index plate.

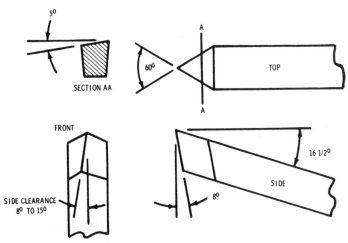

Figure 8-13   Tool for cutting Sharp-V threads.

Thread-Cutting Tools

The shape of the cutting tool depends on the type of thread to be cut. The tool must be ground and set accurately, or the correct thread form will not be tolerated. Various gauges are available that can be used as guides in grinding.

The shape of the cutting tool for cutting a Sharp-V 60° thread is shown in Figure 8-13. Sometimes the point is not ground off, and the thread is cut with the Sharp-V bottom (obsolete); this should never be done when maximum strength is desired.

The Acme screw thread form is often found in power transmissions where heavy loads necessitate close-fitting threads. Another common application for Acme threads is in the lead screws and feed screws of precision machine tools. The cutting tool for cutting external Acme threads is shown in Figure 8-14, and the cutting tool for internal Acme threads is shown in Figure 8-15.

The square thread form is used for many vise and clamp screws. In cutting a square thread with a large lead, the

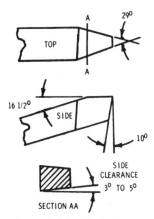

Figure 8-14   Tool for cutting an external Acme thread.

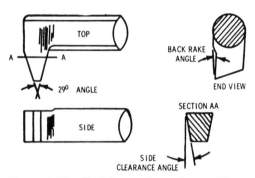

Figure 8-15 Tool for cutting an internal Acme thread.

tool angles must be absolutely correct. Clearance should be allowed on two sides, tapering from both the top and front of the tool. The tool must be fed directly into the work with the cross feed (or compound-rest feed), and care must be taken to avoid chatter and "hogging in." The tool for cutting external square threads is shown in Figure 8-16, and the tool for cutting internal square threads is shown in Figure 8-17.

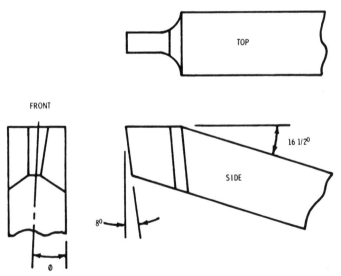

Figure 8-16 Tool for cutting external square threads.

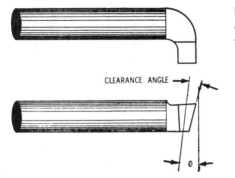

Figure 8-17 Tool for cutting internal square threads.

The Whitworth thread form has been used mostly in England. Great care is required in grinding this cutting tool, so that it will produce the correct radius at both the top and bottom of the screw thread.

Thread-Cutting Operations

The Sharp-V thread is used as an example for thread cutting. If a change-gear lathe is used, the change gears should be properly in mesh, and the operator should note the following:

- For right-hand threads, the carriage must travel toward the headstock.
- For left-hand threads, the carriage must travel toward the tailstock.
- Reversing the direction of rotation of the lead screw (by shifting the tumbler gears) will cause the carriage to move in the opposite direction, without changing the direction of spindle rotation.

Preliminary Setup for Thread Cutting

If a lathe with a quick-change gearbox is used, it is necessary only to place the tumbler levers in the proper position to cut the desired number of threads per inch, as indicated on the index plate. In addition to setting up the lathe to cut the desired number of threads per inch, the cutting tool and the work must be mounted properly.

Setting the Cutting Tool

For cutting external threads, the top of the threading tool should be placed exactly on center (Figure 8-18). The tool should be set

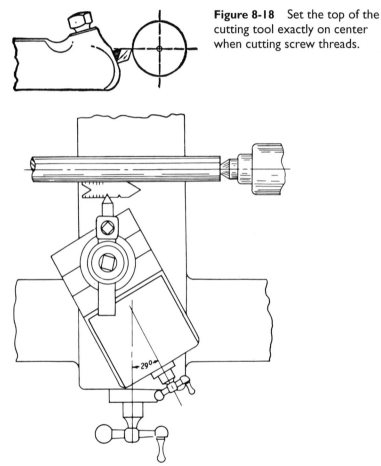

Figure 8-18 Set the top of the cutting tool exactly on center when cutting screw threads.

Figure 8-19 Setup for cutting a 60° American National standard thread. *(Courtesy Cincinnati Milacron Co.)*

square with the work (Figure 8-19). The center gauge can be used to adjust the point of the cutting tool so that the angle of the thread will be correct.

For cutting internal threads, the cutting tool should also be set exactly on center and square with the work (Figure 8-20). The boring bar should be as large in diameter and as short in length as possible to prevent springing. Allow sufficient clearance between the threading tool and the inside diameter of the hole to permit backing out the

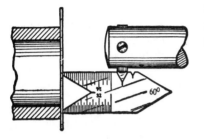

Figure 8-20 Set the cutting tool square with the work when cutting internal screw threads.

tool when the end of the cut has been reached. More front clearance is required in cutting internal threads, to prevent the heel of the cutting tool from rubbing, than when cutting external threads.

Setting the Compound Rest

For cutting 60° screw threads, the compound rest should be set at an angle of 29° (see Figure 8-19). The depth of cut can be adjusted on the compound-rest micrometer dial. Most of the metal is removed by the left-hand side of the threading tool when the compound rest is set at an angle of 29° (Figure 8-21). This permits the chip to curl out of the way and prevents tearing the thread. Because the angle on the side of the threading tool is 30°, the right-hand side of the threading tool will shave the thread smoothly, leaving a fine finish, although it does not remove enough metal to interfere with the main chip (which is removed by the left-hand side of the threading tool).

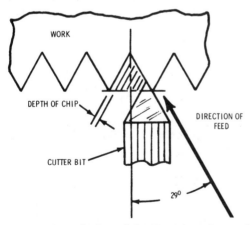

Figure 8-21 Action of the thread-cutting tool when the compound rest is set at a 29° angle.

Cutting Oil for Thread Cutting

Either lard oil or machine oil should be used if a smooth thread is desired on steel. The oil should be applied generously before each cut to prevent tearing the steel by the cutting tool, causing a rough finish on the sides of the threads.

Cutting the Threads

After the lathe has been set to cut the desired number of threads per inch, and both the thread-cutting tool and work have been mounted properly, a very light trial cut should be taken (Figure 8-22). The chief purpose of the trial cut is to make certain that the desired thread pitch is being obtained.

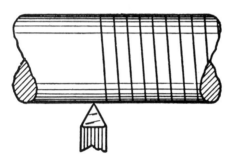

Figure 8-22 A trial cut can be used to check the lathe setup for the correct thread.

The number of threads per inch can be checked by placing a steel rule against the work so that the end of the rule rests on the point of a thread or one of the scribed lines (Figure 8-23). The

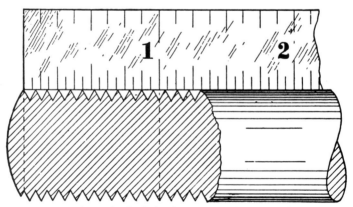

Figure 8-23 A method of checking screw thread pitch.

number of spaces between the end of the rule and the first inch mark indicates the number of threads per inch (11½ threads per inch in Figure 8-23).

Use of Thread Stop

The thread stop can be used to relocate the cutting tool for each successive cut (Figure 8-24). The thread-cutting tool must be withdrawn quickly after each cut to prevent the point of the tool from digging into the metal. It is difficult to stop the threading tool abruptly, so provision for clearance is usually made at the end of the cut. A neck or groove provides clearance for the tool to run out at the end of the cut.

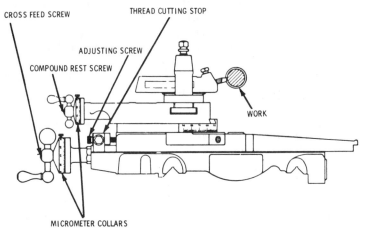

CROSS FEED SCREW

THREAD CUTTING STOP

ADJUSTING SCREW

COMPOUND REST SCREW

WORK

MICROMETER COLLARS

Figure 8-24 A thread stop attached to the dovetail of the saddle.
(Courtesy South Bend Lathe, Inc.)

The point of the cutting tool should be set so that it touches the work lightly. Then, the thread stop should be locked to the saddle dovetail at about ¼ (0.25) inch from the base of the compound rest. The thread stop adjusting screw should be turned until the shoulder is tight against the stop.

To withdraw the cutting tool after each successive cut, turn the cross-feed handle several turns to the left, and return the carriage to the point where the thread is to begin. Then, turn the cross-feed handle to the right until the thread-cutting stop screw contacts the thread stop. This places the cutting tool in its original position. Turn inward 0.002 inch or 0.003 inch on the compound-rest micrometer collar for each successive cut.

Use of the Thread Dial Indicator

The thread dial indicates the relative positions of the lead screw, the carriage, and the spindle of the lathe (Figure 8-25). This permits dis-

engaging the half-nuts at the end of a cut, returning the carriage quickly to the starting point by hand, and re-engaging the half-nuts at the correct point to ensure that the cutting tool is following exactly in the original cut.

When the gear on the lower end of the thread dial indicator meshes with the lead screw, any movement of the carriage or lead screw is indicated by a corresponding movement on the graduated dial at the top. The points of half-nut engagement will vary with the individual lathe. For example, on the thread dial indicator shown in Figure 8-25, the points at which the half-nuts may be engaged for successive cuts will be indicated as follows:

Figure 8-25   Thread dial indicator attached to the lathe carriage.
(Courtesy South Bend Lathe, Inc.)

- For *even-numbered threads*, engage the half-nuts at any line on the dial, or each ⅛ revolution.
- For *odd-numbered threads*, engage the half-nuts at any numbered line on the dial or each ¼ revolution.
- For *half-numbered threads* (11½ threads per inch), close the half-nuts at any odd-numbered line, or each ½ revolution.
- For *quarter-numbered threads* (4¾ threads per inch), return to the original starting point for each successive cut.

Resetting the Threading Tool

If it is necessary to remove the threading tool before the thread has been completed, the tool must be carefully adjusted to follow the original groove when it is replaced in the lathe. All lost motion can be taken up by rotating the drive belt by hand. By adjusting the compound-rest and the cross-feed screws simultaneously, the threading tool can be adjusted to enter the original groove.

Fitting and Checking Threads

A screw thread pitch gauge can be used to check the finer pitches of threads (Figure 8-26). Either the nut that is to be used on the thread or a ring type of thread gauge can be used to check the thread

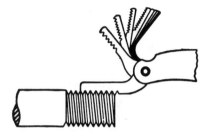

Figure 8-26 A screw thread pitch gauge. *(Courtesy South Bend Lathe, Inc.)*

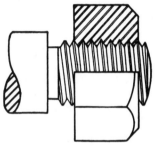

Figure 8-27 Fitting a screw thread to a nut to check fit. *(Courtesy South Bend Lathe, Inc.)*

(Figure 8-27). The nut should fit snugly without any play, but it should not bind on the thread at any point.

Finishing the End of a Thread

A 45° chamfer is commonly used to finish the end of a thread on bolts, cap screws, and so on (Figure 8-28). A forming tool is often used to round the ends of machine parts and special screws (Figure 8-29).

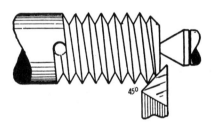

45°

Figure 8-28 A 45° chamfer is commonly used to finish the end of the screw thread on bolts, cap screws, and so on.

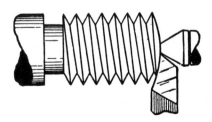

Figure 8-29 Finishing the end of a thread by rounding with a forming tool.

Summary
The screw thread is a very important mechanical device used for adjusting, fastening, and transmitting motion. Thread cutting with the engine lathe requires a thorough knowledge of thread-cutting principles and procedures.

The most common form of screw thread is one that has symmetrical sides inclined at equal angles to a vertical centerline through the apex of the thread. The American National screw thread was the standard thread used in the United States for many years, but now has been replaced by the Unified, Whitworth, and Acme forms. In the Unified system, the crest of the external thread can be either flat (preferred by American industry) or rounded (preferred by British industry). The crest is flat on internal threads. Otherwise, the Unified thread is practically the American National form long used in the United States.

Diameter pitch combinations are grouped to form thread series that are distinguished by the number of threads per inch applied to a specific diameter. There are 11 standard series combinations (such as course-thread, fine-thread, and extra-fine thread).

The shape of the cutting tool depends on the type of thread to be cut. The tool must be ground and set accurately, or the correct thread form will not be obtained. Various gauges are available that can be used as guides in grinding. The Whitworth thread form has been used mostly in England. Extreme care is required in grinding this cutting tool so that it will produce the correct radius at both the top and bottom of the screw thread. The Acme screw thread form is often found in power transmissions where heavy loads necessitate close-fitting threads. Another common application for Acme threads is in the lead screws and feed screws of precision machine tools. The square thread form is used for many vise and clamp screws.

Review Questions
1. Name the various screw thread forms.
2. What country uses the Whitworth thread form?
3. What is meant by the angle of thread?
4. What standard thread was used for years in the United States?
5. What is meant by length of engagement?
6. What is now the basic standard screw thread in the U.S.?
7. What screw thread has largely replaced the square thread?

8. What is the Dardelet thread?

9. When and where was the International Metric Thread System adopted?

10. What do the following mean?
  A. UNC
  B. UNF
  C. UNEF

11. Identify the following terms:
  A. Angle of thread
  B. Axis of screw
  C. Base of thread
  D. Basic
  E. Crest
  F. Crest clearance
  G. Depth of engagement
  H. Depth of thread
  I. External thread
  J. Helix angle

12. What is meant by the pitch of a screw thread?

13. Where is the root of a thread located?

14. What is the difference between simple and compound gears?

15. What is compound gearing?

Chapter 9

Lathe Attachments

The many attachments that are available for the engine lathe make it a very versatile machine. Some of these attachments facilitate the usual lathe operations, and others permit a variety of machining operations that are not strictly lathe work.

Work Support Attachments

In many lathe operations, the work cannot be held between the lathe centers, and an attachment must be used to aid in supporting the work. This is especially true when it is necessary to bore, drill, thread, or perform other similar internal operations on the end of a long piece of work.

Follower Rest

The *follower rest* is used to support long slender stock to prevent its springing away from the cutting tool. The follower rest is bolted to the saddle of the carriage. Therefore, it "follows" the cutting tool. The frame of the follower rest is shaped like a question mark, and it has both vertical and horizontal adjustable jaws (Figure 9-1).

The jaws of the follower rest form a true bearing for the work, permitting it to turn without binding. In setting the follower rest jaws, first remove the guard over the cross-feed screw. Place a small piece of paper over the screw to keep off chips during the cutting operation. The dovetail ways should be wiped clean. Adjust the jaws with the carriage near the tailstock after a short portion of the work has been turned. Set the vertical jaw to touch the top of the workpiece. A small piece of cellophane can be placed between the work and the jaws to determine the proper amount of friction. During the cutting operation, plenty of lubricant should be applied to the workpiece where it contacts the jaws of the follower rest. The jaws should be adjusted after each cut to retain accuracy.

Steady Rest

The *steady rest* is used to support one end of the long, slender stock when the other end is centered in a chuck and the operation being performed is such that the tailstock cannot be used (Figure 9-2). The frame of the steady rest is circular in shape and is fastened to the bed ways of the lathe (Figure 9-3). The adjustable jaws form a bearing for the work, holding it in centered position.

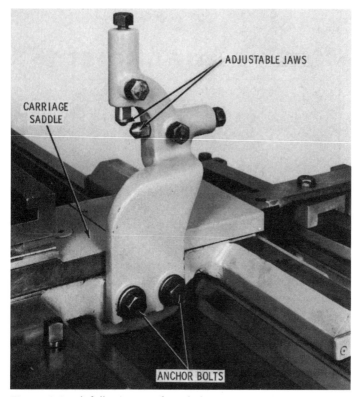

Figure 9-1 A follower rest for a lathe. *(Courtesy Cincinnati Milacron Co.)*

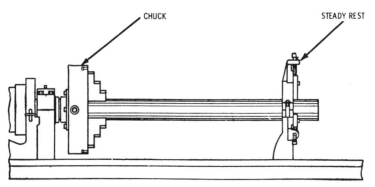

Figure 9-2 Work mounted with one end centered in a chuck and the other end supported by a steady rest.

Figure 9-3 A steady rest for a lathe. *(Courtesy Cincinnati Milacron Co.)*

Positioning the jaws properly is essential when mounting the workpiece. The jaws must form a true bearing for the workpiece, permitting it to turn freely without binding or play. With one end of the work mounted in the chuck, slide the steady rest close to the chuck jaws, tighten the base clamp, adjust the jaws of the steady rest, and lock them in position on the work. A small piece of cellophane placed between the jaws and the work can be an aid in obtaining the proper bearing. Advance each jaw of the steady rest until it barely touches the work, and remove the cellophane. After tightening both the locknut and the clamp screw on each jaw, loosen the base clamp. Slide the steady rest into proper position, and retighten the base clamps. If extreme accuracy is required, trueness of the work should be checked with a dial gauge.

In machining the end of a long cylindrical piece, it is preferable to secure one end of the workpiece in a chuck (see Figure 9-2). However, heavy belt lacing can be bound around a lathe dog to hold it to a faceplate so that the end of the work can be mounted on a headstock center (Figure 9-4). This is a kind of makeshift setup that can be used by a person with insufficient skill for mounting the work in an independent chuck with precision.

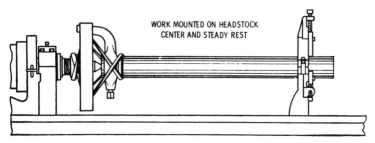

Figure 9-4 Belt lacing can be used to mount the work on a headstock center and steady rest.

Automatic Feed Stops

Automatic devices that disconnect the power feed after a predetermined distance of travel are often used on lathes in production work. These devices can be used to enable a single machinist to operate two or more lathes simultaneously. The operator can place the work in a lathe, engage the power feed for a cut, and proceed to the next lathe. When the end of the cut is reached, the power feed is automatically disconnected.

Carriage Stops

The *micrometer carriage stop* is especially desirable for production work where long longitudinal cuts are required. It is indispensable for accurate control of length of cut. The fixed stop can be located at any position along the bed ways. It can be fixed at any position for repetitive work. The micrometer collar is graduated in thousandths of an inch (Figure 9-5). The micrometer carriage stop can also be used to face shoulders to an exact length.

The *four-position length and depth stop* is designed for production turning and for facing multiple diameters and shoulders (Figure 9-6). This turret-type stop has four individually adjusted stop screws. The

Figure 9-5 A micrometer carriage stop attachment for a lathe.
(Courtesy Cincinnati Milacron Co.)

length stop mounts on the front bed ways, and the depth stop mounts on the carriage wings and bridge. A handy *length-measuring attachment* can be used to measure either the forward or the reverse movement of the lathe carriage within 0.001 inch (Figure 9-7).

Cross-Slide Stop
The *ball-type cross-feed stop* enables the operator to retract the cross slide at the end of a cut and to return quickly to its original setting prior to the next cut (Figure 9-8). When the thumbscrew is

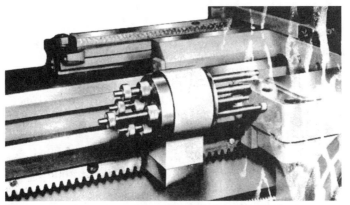

Figure 9-6 A four-position length and depth stop attachment.
(Courtesy Cincinnati Milacron Co.)

Figure 9-7 A length-measuring attachment used to measure the forward or reverse carriage movement to 0.001 inch. *(Courtesy Cincinnati Milacron Co.)*

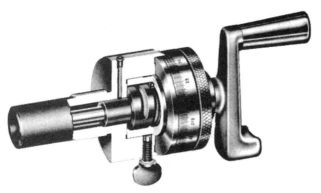

Figure 9-8 Cross-feed stop. *(Courtesy Cincinnati Milacron Co.)*

loose, the cross-feed screw can be used normally for full length of travel of the cross slide. The stop can be set quickly by adjustment of the thumbscrew.

Other Attachments

A number of machining attachments are available to increase the capability of the lathe for performing a variety of operations.

Taper Attachment

A typical *taper* attachment for a lathe is shown in Figure 9-9. This is probably the most widely used method of machining tapers on

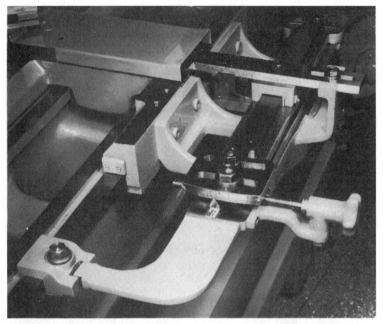

Figure 9-9 A lathe taper attachment. *(Courtesy Cincinnati Milacron Co.)*

the engine lathe. It is much preferred to the tailstock set-over method for machining large tapers. Tapers up to 4 inches per foot (20°) can be machined with the taper attachment.

Most taper attachments consist of a bar at the back of the lathe. The bar is engaged by the cross slide, which moves the cutting tool transversely closer to or farther from the lathe center axis as the cutting tool travels longitudinally along the lathe bed. Duplicate tapers can be cut quickly on pieces of different lengths. Taper boring, which is impossible with the tailstock set over, can be performed easily with the taper attachment.

The *swivel taper bar* (*L* in Figure 9-10) is graduated in both degrees *J* and inches per foot *N*, and represents the included angle or taper. The taper bar moves only one-half the amount indicated on the scale.

In adjusting the taper attachment (see Figure 9-10), the cross-slide bar *M* should be kept centrally located on the swivel taper bar for easy adjustment (keep the screw *F* in the cross-slide bar over the stud *K* in the taper bar). To set the taper attachment for a desired taper, loosen the clamping nuts *C*, *D*, *E*, *F*, and *G*. Then, the swivel-taper

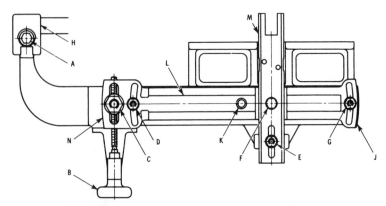

Figure 9-10   Diagram of a typical taper attachment showing parts.
(Courtesy Cincinnati Milacron Co.)

bar can be adjusted to the approximate setting by hand. A final adjustment can be made by tightening the nut C and using the adjusting screw B. Position the cutting tool so that the taper attachment will travel the full length of the part to be turned. Then, tighten the engaging nut A to engage the taper attachment.

The attachment can be disengaged by loosening the engaging nut A and tightening the clamping nut E, which must remain tight except during the tapering operation. The clamping block H should be removed if the taper attachment is not used frequently, to reduce wear on the bed ways.

The taper attachment can also be used to bore a tapered hole. Set the attachment in the same manner as for turning external tapers, except that the angle of the taper bar is usually reversed. Make certain that the tool will clear the smallest diameter of the tapered hole and that the tool is set on center to avoid boring either a concave or a convex surface onto the workpiece.

Tracer Attachment

The *tracer* attachment for the lathe can be used for quick and accurate duplication of parts (Figure 9-11). It can be set to cut taper angles between 0° and 90° in both facing and turning operations. The template is positioned on the longitudinal template support bar.

The hydraulic power slide has a normal movement angle of 45° to the centerline of the lathe. Maximum power slide travel is 3¾ (3.750) inches. This corresponds to a movement of 2⅝ (2.625) inches perpendicular to or parallel to the centerline of the work.

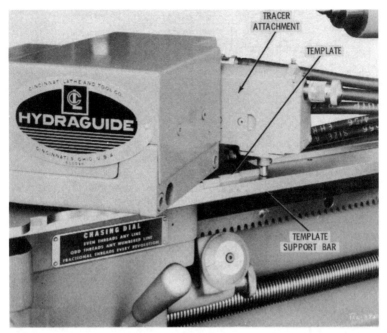

Figure 9-11 Tracer attachment for a lathe. *(Courtesy Cincinnati Milacron Co.)*

Milling Attachment

This attachment can be used for milling work in a small shop that cannot afford to install a milling machine (Figure 9-12). To install the milling attachment on the lathe, remove the compound rest, and clamp the base of the milling attachment in its place. The milling attachment equips the engine lathe for face milling, cutting keyways and slots, milling dovetails, squaring shafts, and making dies, molds, and so on.

The vise position is controlled by a feed screw with a graduated micrometer collar. The attachment can be swiveled to hold work at any angle.

The cutting speed for a milling operation should be approximately two-thirds the speed recommended for turning the material in the lathe. The cut is controlled by the lathe carriage handwheel, the lathe cross-feed screw, and the vertical adjusting screw at the top of the milling attachment. All milling cuts should be taken with the direction of cutter rotation against the direction of feed of the workpiece.

Figure 9-12   Milling attachment for a lathe. *(Courtesy Atlas Press Co.)*

Gear-Cutting Attachment

Spur and bevel gears of all kinds can be cut on the lathe *gear-cutting* attachment. The attachment equips the lathe for graduating and milling, external key-seat cuts, angle cuts, spline cuts, slotting work, and regular dividing head milling work.

The device is designed for mounting on the milling vise. The cutter is mounted on a mandrel that turns on the lathe centers.

Grinding Attachment

A tool-post grinder can be mounted in the tool slide of the lathe compound rest (Figure 9-13). This attachment equips the lathe for both internal and external precision grinding (Figure 9-14). It can be used to grind reamers and milling cutters, dies, gauges, bushings, bearings, shafts, valves, and valve seats. The grinding attachments are driven by small electric motors; the lathe spindle must be reversed for grinding operations (see Figure 9-13), and as the grinders are constructed for attachment to the compound rest, the grinding wheel can be adjusted to the proper position for the various grinding operations. As grinding is a finishing operation, the work should be turned as near the finished sizes as possible before beginning the operation.

Turret Attachments

Turrets can be used on the lathe bed, tailstock, and tool post. Advantages of these attachments are elimination of second-operation setups and rapid and accurate machining of duplicate parts on a production basis.

Bed Turret

The bed turret is a cylinder or head arranged to turn and slide on the lathe bed ways (Figure 9-15). It can be fitted with various tools for boring, reaming, tapping, and so on. The head indexes ⅙ of a turn with each complete backward movement of the handwheel.

Figure 9-13 Tool-post grinding attachment for a lathe. *(Courtesy Atlas Press Co.)*

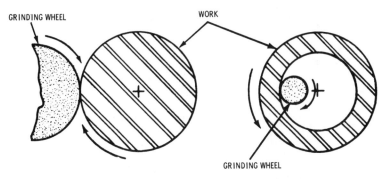

EXTERNAL GRINDING   INTERNAL GRINDING

Figure 9-14 Diagram illustrating external (left) and internal (right) grinding on the lathe. The work must always turn in a direction opposite that of the grinding wheel. For external grinding, it is necessary to reverse the lathe spindle. For internal grinding, the lathe spindle should be run in the forward direction.

Tailstock Turret

This device mounts in the lathe tailstock spindle (Figure 9-16). Any of the six tool stations can be selected quickly by a convenient trip-lever control. The attachment can be used to speed

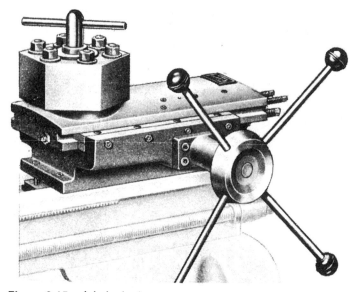

Figure 9-15 A lathe bed turret. *(Courtesy Atlas Press Co.)*

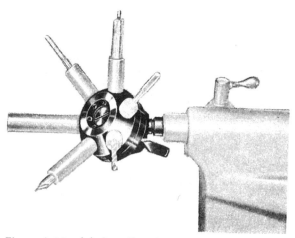

Figure 9-16 A lathe tailstock turret. *(Courtesy Atlas Press Co.)*

short-run work and to eliminate the need for second-operation setups.

Tool-Post Turret

This attachment can be mounted on the compound rest (Figure 9-17). Various cutting tools can be mounted in the turret for facing, turning, and so on. A turn of the handle releases the head for rotating to the next operating position.

Figure 9-17 A tool-post turret. *(Courtesy Atlas Press Co.)*

Summary

Many lathe attachments are available that make the lathe a very versatile machine. In many lathe operations, the work cannot be held between the lathe centers, and an attachment must be used to aid in supporting the steady rest. The follower rest prevents long slender stock from springing away from the cutting tool. The steady rest supports stock when the other end is centered in a lathe chuck and the tailstock cannot be used.

Automatic devices that disconnect the power feed after a predetermined distance of travel are often used on lathes in production work. With this device, a single machinist can generally operate two lathes simultaneously. He can place the work on one lathe, engage the power feed for a cut, and proceed to the next lathe. When the end of the cut is reached, the power feed is disconnected automatically.

The micrometer carriage stop, especially desirable for production work where long longitudinal cuts are required, is indispensable for accurate control of length of cut. The fixed stop can be located at any position along the bed ways and can be fixed at any position for repetitive work. The ball-type cross-feed stop enables the operator to retract the cross slide at the end of a cut and to return quickly to its original setting prior to the next cut.

Taper attachments for a lathe are probably the most widely used method of machining tapers on the engine lathe. They are much preferred to the tailstock set-over method for machining large tapers. Most taper attachments consist of a bar at the back of the lathe.

The tracer attachment for the lathe can be used for quick and accurate duplication of parts. The milling attachment can be used for milling work in a small shop that cannot afford to install a milling machine. Spur and bevel gears of all kinds can be cut on the lathe gear-cutting attachment. A tool-post grinder can be mounted in the tool slide of the lathe compound rest. Turrets can be used on the lathe bed, tailstock, and tool post. There are three types of these attachments: the bed turret, the tailstock turret, and the tool-post turret.

Review Questions

1. What is a follower rest?
2. What is a steady rest?
3. What other attachments are available for various lathe operations?
4. What is an automatic feed stop?
5. Where is the micrometer carriage stop used in production work?
6. What is the purpose of the ball-type cross-feed stop?
7. Where is belt lacing used, and why?
8. What is taper?
9. How are tapers cut on a lathe?
10. What is a tracer attachment?

Chapter 10

Turret Lathes

The turret lathe is a comparatively modern machine that has been developed from the engine lathe by the addition of revolving tool-holding devices called *turrets*. The turret (or head) can be arranged to turn on an upright axis and to slide on the ways of the lathe. It is fitted with holes that hold cutting tools, any of which can be applied as needed in the axial line of the work by turning the head.

At first, turrets were made in circular form and were rotated on a vertical pivot that could be held in any desired position by a binding nut. The periphery of the circular turret was drilled and reamed to hold four tools, which projected from the turret at 90° angles with each other. Later, the turret was made in hexagonal form, and the number of tools held in the turret was increased to six.

The earlier turrets were mounted on the lathe carriage in place of the tool block and were aligned with the lathe axis by means of the cross-feed screw. Lateral feed was obtained from the feed mechanism in the apron attached to the carriage.

The chief objective accomplished by addition of the turret to the earlier lathes was that it enabled the operator to perform drilling, reaming, counterboring, and similar operations on a piece of work without a change of tools other than revolving the turret. After they had been set and adjusted, the tools in the turret required no further alterations as the workpieces were completed and removed from the chuck, and other pieces substituted for a like series of operations. Thus, work could be performed more rapidly by this means. The turret mechanism was developed further by the addition to the number of tools it could carry, by a ratchet arrangement for revolving the turret, by an index plate for holding the turret in any desired position, and by various other improvements in design.

All these improvements led to the design of special lathes in which the improved turret was a special feature, and led to the development of the turret lathe as it is built today.

The turret lathe is one of the most useful machines in the machine shop for the production (in large quantities) of work that requires repeated operations (such as turning, boring, or reaming). The turret is equipped with the appropriate tools, and the work, such as a casting or forging, is held by a chuck.

Classification

Turret lathes are classified as either *horizontal* or *vertical*. In general, unless specified, a turret lathe is understood to be a horizontal turret lathe. Horizontal turret lathes are classified as either *bar machines* or *chucking machines*.

Bar machines are used to machine bar stock or castings that are about the same size and shape as bar stock. The chucking machines are used to machine castings and forgings that must be held in chucks or fixtures. Bar and chucking machines are classified further as either *ram-type turret lathes* or *saddle-type turret lathes*.

Ram Type

On the ram-type turret lathe, the turret is mounted on a slide (or ram) that moves back and forth on a saddle clamped to the lathe bed (Figure 10-1). This machine is a fast and easily operated machine for small work. It is quickly and easily operated because the hexagonal turret can be moved back and forth without moving the entire saddle unit. The machine has a short turret stroke and an automatic index on the turret. The ram-type turret lathe is well suited for bar work and chucking work where the overhang of the ram can be kept short.

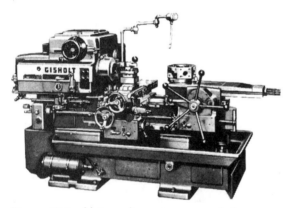

Figure 10-1   Universal ram-type turret lathe. *(Courtesy Gisholt Machine Co.)*

Saddle Type

On the saddle-type turret lathe, the turret is also mounted on the saddle, but the entire saddle unit moves back and forth on the lathe bed (Figure 10-2). This machine is more heavily constructed than the ram type and is more suitable for work requiring long turning and boring cuts. It has a long turret stroke and a rigid turret mounting.

Figure 10-2   Saddle-type turret lathe. *(Courtesy Gisholt Machine Co.)*

Basic Construction

The basic parts of a turret lathe are headstock, ram saddle, turret, cross-slide carriage, and chuck. In addition, various automatic devices can be used on turret lathes.

Headstock

One of the most important operating units of any turret lathe is the headstock. Either an electric head with a multiple-speed motor mounted directly on the spindle or an all-geared head is used on the modern turret lathe. These heads provide a wider range of speeds and permit heavier cuts than were possible on the earlier cone-drive heads. The operator needs only to set the dial to the diameter of the work (or the desired speed), and the spindle speed selector will automatically shift to the correct speed.

Turret Ram and Saddle

In the ram-type turret lathes, the ram provides a base for the turret (Figure 10-3). The turret is automatically unclamped and indexed by backward movement of the ram. Forward movement clamps the turret before it leaves the saddle. The turret is free for backing up or skip indexing at one point in the travel of the ram. The longitudinal feed handwheel moves the ram in the saddle and permits leverage when feeding by hand from a normal working position.

Following are the three general classes of rams based on the movement or movements of the turret:

- *Plain*—The plain slide (or ram) provides only longitudinal movement. The turret is rotated automatically to bring the next tool into position on the backward or return stroke.

Figure 10-3 Turret ram and saddle for a turret lathe. *(Courtesy Gisholt Machine Co.)*

- *Offset*—The offset slide (or set-over slide) provides both longitudinal and transverse movements. The transverse motion can be used for radial facing or recessing with a single-point tool.
- *Universal*—The universal slide (or ram) has both the longitudinal and transverse movements. In addition, it has an arrangement consisting of an intermediate plate between the cross slide and ram to permit swiveling for such work as turning or boring of tapered surfaces.

These three types of rams are shown in Figure 10-4. The saddle-type turret lathe is a form of universal machine in which the turret is mounted on a swivel placed on the saddle (Figure 10-5).

Turret

The hexagon-shaped turret in the ram-type turret lathe is mounted on a slide, or ram, which moves back and forth in the saddle (see Figure 10-3). The saddle is clamped to the bed ways in proper position for the job to be performed.

Square turrets on ram-type turret lathes are adjustable along the slide to obtain proper tool clearances. The square turret is mounted directly on the slide in saddle-type turret lathes.

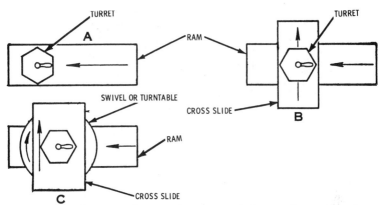

Figure 10-4   Diagram showing three general classes of rams found on turret lathes: (A) plain, (B) offset, and (C) universal.

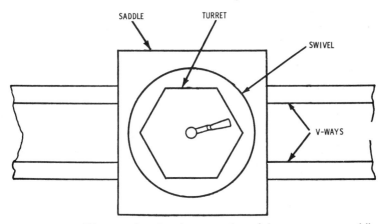

Figure 10-5   Diagram showing the mounting of the turret on a saddle-type turret lathe.

Cross-Slide Carriage

The *cross-slide carriage* is held to the bed ways by flat hold-down plates with tapered gibs for wear adjustment (Figure 10-6). Eight selective, reversible power cross and longitudinal feeds are available on the cross-slide carriage shown in Figure 10-6.

A dial-type feed selector gives a direct reading in thousandths per revolution for each of the eight power feeds. A lever at the right of the apron engages the cross feed, and a lever at the lower-left front of the apron selects either forward or reverse feed.

Figure 10-6 Cross-slide carriage on a ram-type turret lathe.
(Courtesy Gisholt Machine Co.)

Chuck

A *collet chuck* (Figure 10-7) is commonly found on turret lathes. It is hydraulically operated and controlled by movement of the "chuck-unchuck" lever. Round, square, or hexagon bar stock can be handled easily with this type of chuck.

Figure 10-7 A hydraulically operated collet chuck. It is controlled by movement of the "chuck-unchuck" lever.
(Courtesy Gisholt Machine Co.)

Turret Lathe Attachments

Numerous automatic devices are used on the modern turret lathe to control turret movements that bring the tool into position at the correct moment and to feed the work through the chuck after each piece is finished. Various "stops" are placed on top of the lathe bed—not to disengage the feed, but to arrest the travel of the carriage while the feed continues to hold the carriage firmly against the stop.

A *four-position stop roll* (Figure 10-8) can be used with both front and rear cross-slide rolls. The stop roll is

Figure 10-8 Four-position stop-roll attachment.
(Courtesy Gisholt Machine Co.)

hand-indexed, and it can be set up quickly to trip cross feeds in either direction. Each feed trip dog is equipped with an independently adjustable dead stop to permit hand feeding to an exact size, while referring to a dial or indicator, after the power feed has been disengaged. Also, the dead stop can be used for accurate setting of the cross slide for turning operations.

A *taper attachment* for the universal carriage is a separate tool bolted to the T-slots on the rear of the cross slide (Figure 10-9). The rear tool block can be used for taper turning without affecting normal use of the square turret at the front of the cross slide.

A *rapid-traverse cross-slide attachment* (Figure 10-10) reduces operator effort and speeds machining operations. An independent electric motor drive unit is mounted on a bracket at the rear of the cross slide and connected by gears to the cross-feed screw. Manual operation of a switch control provides traverse in either direction.

Superprecision Turret Lathes

The bench turret lathe operates in much the same manner as the larger turret lathes. This type of machine tool is designed for short-run and long-run superprecision secondary machining operations that require extremely close tolerances.

Figures 10-11 through 10-39 illustrate the bench turret lathe, as well as a wide variety of features, attachments, and operations performed on this precision machine tool.

Figure 10-9   Taper attachment for the universal carriage.
(Courtesy Gisholt Machine Co.)

Figure 10-10   An electric rapid-traverse attachment to the cross slide of a turret lathe. *(Courtesy Gisholt Machine Co.)*

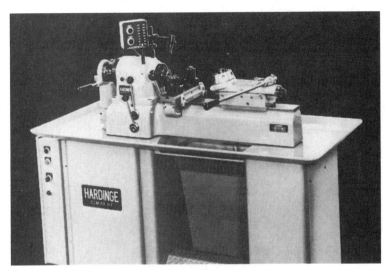

Figure 10-11 This bench turret lathe is a superprecision second-operation machine for machining extremely close tolerances.

(Courtesy Hardinge Brothers, Inc.)

Figure 10-12 A drilling operation. *(Courtesy Hardinge Brothers, Inc.)*

Figure 10-13 An internal turning operation. *(Courtesy Hardinge Brothers, Inc.)*

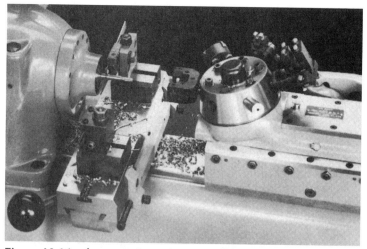

Figure 10-14 An operation using a cutoff tool. *(Courtesy Hardinge Brothers, Inc.)*

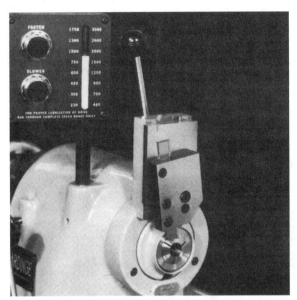

Figure 10-15 Vertical cutoff slide. *(Courtesy Hardinge Brothers, Inc.)*

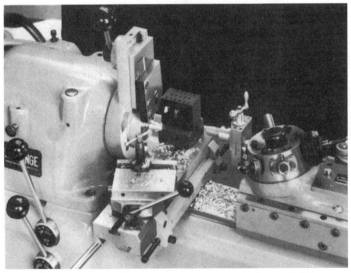

Figure 10-16 A cutoff operation using a vertical cutoff slide.
(Courtesy Hardinge Brothers, Inc.)

Figure 10-17   A boring operation involving a tapered hole.

(Courtesy Hardinge Brothers, Inc.)

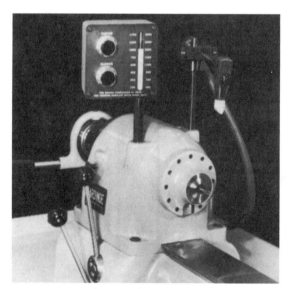

Figure 10-18   A collet for holding parts.

(Courtesy Hardinge Brothers, Inc.)

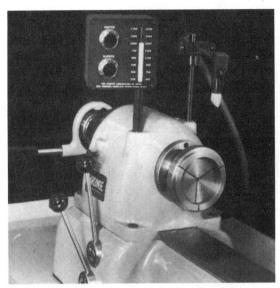

Figure 10-19 A step chuck for holding parts.
(Courtesy Hardinge Brothers, Inc.)

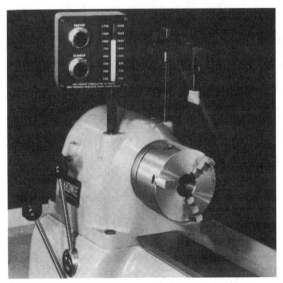

Figure 10-20 A three-jaw chuck for holding parts.
(Courtesy Hardinge Brothers, Inc.)

Figure 10-21 An internal chucking operation.
(Courtesy Hardinge Brothers, Inc.)

Figure 10-22 "Slower" button is pressed to decrease the spindle speed. *(Courtesy Hardinge Brothers, Inc.)*

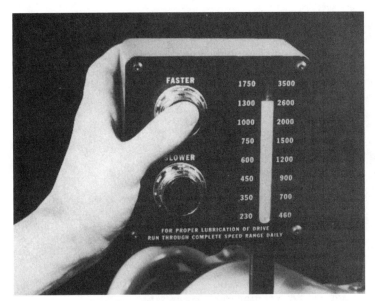

Figure 10-23   "Faster" button is pressed to increase the spindle speed. *(Courtesy Hardinge Brothers, Inc.)*

Figure 10-24   The double-tool cross slide. *(Courtesy Hardinge Brothers, Inc.)*

Figure 10-25 Preloaded ball-bearing turret, specially designed for speed and ease of operation. *(Courtesy Hardinge Brothers, Inc.)*

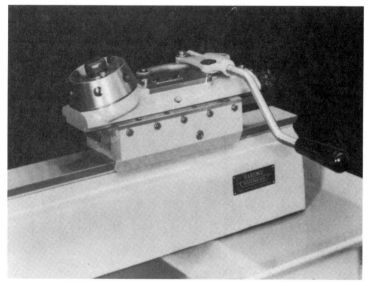

Figure 10-26 A turning operation using a step chuck.
(Courtesy Hardinge Brothers, Inc.)

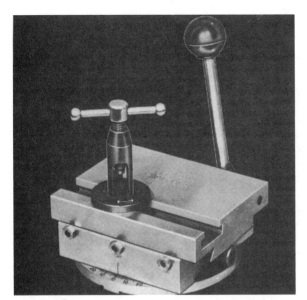

Figure 10-27 Straight and taper turning slide for double-tool cross slide. *(Courtesy Hardinge Brothers, Inc.)*

Figure 10-28 High-production precision turning of stainless steel carburetor needle valves. *(Courtesy Hardinge Brothers, Inc.)*

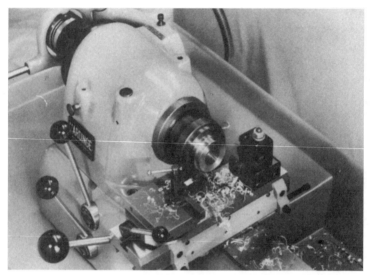

Figure 10-29   Slide set for straight turning of a large-diameter aluminum part. *(Courtesy Hardinge Brothers, Inc.)*

Figure 10-30   Slide setup in front position of the double-tool cross slide for precision turning of a taper on a brass part. *(Courtesy Hardinge Brothers, Inc.)*

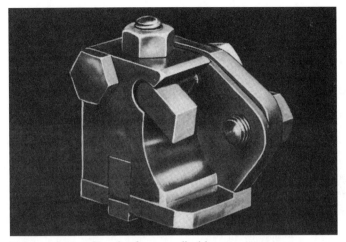

Figure 10-31 Circular form toolholder. *(Courtesy Hardinge Brothers, Inc.)*

Figure 10-32 The holder fits directly to the double-tool cross-slide toolholder block. *(Courtesy Hardinge Brothers, Inc.)*

Figure 10-33 Multiple toolholder. *(Courtesy Hardinge Brothers, Inc.)*

Figure 10-34 With the multiple toolholder, many operations such as undercutting, chamfering, and grooving can be done in one operation.
(Courtesy Hardinge Brothers, Inc.)

Figure 10-35   Cutoff toolholder. *(Courtesy Hardinge Brothers, Inc.)*

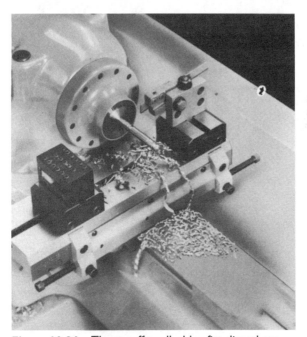

Figure 10-36   The cutoff toolholder fits directly to
the front or rear tool block on the double-tool cross slide.
(Courtesy Hardinge Brothers, Inc.)

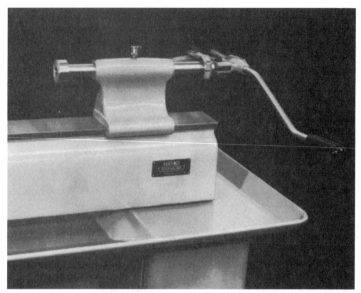

Figure 10-37 This slide is for deep hole drilling, long threading, box tool turning, lapping, or any other operation requiring tool travel up to 5½ inches. *(Courtesy Hardinge Brothers, Inc.)*

Figure 10-38 Deep hole drilling. *(Courtesy Hardinge Brothers, Inc.)*

Figure 10-39   Production threading. *(Courtesy Hardinge Brothers, Inc.)*

Summary

The turret lathe is a comparatively modern machine that has been developed from the engine lathe by the addition of revolving tool-holding devices called turrets. The turret or head can be arranged to turn on an upright axis and to slide on the ways of the lathe; it is fitted with holes that hold cutting tools, any of which can be applied as needed in the axial line of the work by turning the head.

Turret lathes are classified as either horizontal or vertical. In general, a turret lathe is understood to be a horizontal type. Turret lathes are usually classified as either bar machines or chucking machines. The chucking machine is used to machine castings and forgings that must be held in chucks or fixtures. The bar machine is used to machine bar stock or castings that are about the same size and shape as bar stock. Bar and chucking machines are further classified as ram-type turret or saddle-type turret lathes.

On the ram-type lathe, the turret is mounted on a slide, or ram, which moves back and forth on a saddle that is clamped to the lathe bed. On the saddle lathe, the turret is also mounted on the saddle. However, the entire saddle unit moves back and forth on the lathe bed.

The basic parts of a turret lathe are headstock, ram saddle, turret, slide carriage, and chuck. Various automatic devices can be used on the turret lathe, such as stops placed on top of the lathe bed to limit the travel of the carriage while the feed continues to hold the carriage firmly against the stop.

A collet chuck is commonly found on turret lathes. It is hydraulically operated and controlled by movement of the "chuck-unchuck" lever. Round, square, or hexagon bar stock can be handled easily with this type of chuck.

Numerous automatic devices are used on the modern turret lathe to control turret movement that bring the tool into position at the correct movement and to feed the work through the chuck after each piece is finished.

Review Questions

1. How are turret lathes classified?

2. What is a chucking machine?

3. What is a bar machine?

4. What are the advantages of a turret lathe over an engine lathe?

5. What are the three general classes of rams based on movement?

6. How is the cross-slide carriage held to the bed ways?

7. What is a rapid-traverse cross-slide attachment used for?

8. What is a collet used for?

9. Describe an internal chucking operation.

10. What is needed in order to perform such operations as undercutting, chamfering, and grooving?

Chapter 11

Milling Machines

A milling machine is a power-driven machine that cuts metal by means of a multitooth rotating cutter. The machine is constructed in such a manner that the workpiece is fed to a rotary cutter—instead of revolving as on a lathe, or reciprocating as on a planer.

Adaptation

A variety of operations can be performed on the milling machine. A multitooth milling cutter remains sharp much longer than a single cutting tool. The cutting action of the milling machine is continuous, as compared to the intermittent cutting action of the shaper and planer. Therefore, many kinds of workpieces can be machined more economically on the milling machine than on a shaper or planer.

Milling machines are adapted for production of workpieces having intricate profiles that must be interchangeable. Large castings or forgings can be handled on the milling machine. Many specialized types of milling machines have been developed, which has broadened considerably the range of application of the milling machine.

Basic Construction and Classification

The milling machine has a power-driven spindle. An arbor for holding multitooth cutters fits into the spindle. The cutting edges or teeth on revolving circular cutters remove a controlled amount of metal at each revolution of the cutter. The workpiece is mounted on a movable table and is fed against the cutter. The table can be moved either by hand-feed or by power-feed. When several cutters are mounted on the arbor, several surfaces can be machined in one operation. The knee-and-column type of milling machine is commonly found in industry (Figure 11-1).

Knee-and-Column

The knee-and-column types are usually cast in one piece to make up the main casting of the milling machine. The column is thick-walled and strongly braced to support the other parts of the machine, and usually includes a tank for coolant used on the cutter. The inner space of the column houses the driving motor and the gear mechanism for transmitting power to the spindle and table of the milling machine.

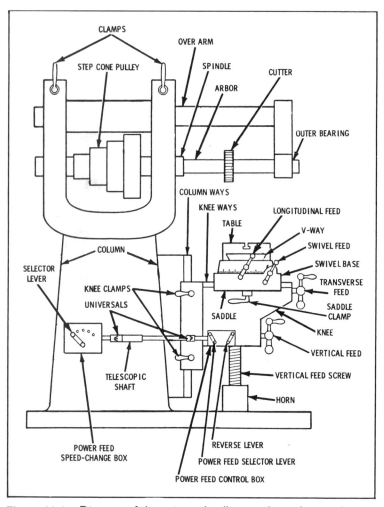

Figure 11-1 Diagram of the universal milling machine, showing knee-and-column design, basic parts, and controls.

The top of the column supports the overarm (or overarms). The design of overarms depends on the manufacturer of the machine. The arbor support slides on the overarm.

The arbor support clamps onto the overarm and supports the milling arbor. The bearing in the lower end of the arbor support should be in perfect alignment with the spindle axis. The *inner support* supports the arbor near the middle. A bearing sleeve placed between the collars on the arbor fits into the large hole. A smaller hole in the *outer support* holds the ground end of the arbor.

The spindle is hollow throughout its entire length. It revolves in bearings in the upper end of the column; the front end has a tapered hole for receiving the standard shanks of milling arbors (Figure 11-2).

Figure 11-2   Spindle drive gearing of a milling machine. *(Courtesy Cincinnati Milacron Co.)*

On the face of the column, a wide slide (usually of dovetail design) is provided to ensure proper alignment of a sliding knee or bracket. The knee can be raised and lowered on the column to adjust the depth of cut for jobs of various sizes. The knee is a casting with two of its sides machined at right angles. The vertical machined side of the knee slides on the ways on the face of the column. The top of the knee is machined at a right angle to the vertical surface and supports the saddle, which slides on the knee either toward or away from the face of the column. A sturdy elevating screw raises and lowers the knee.

The saddle is mounted at the top of the knee and supports the table. A precisely machined surface (usually dovetails) on the top side of the saddle holds the milling machine table (Figure 11-3).

The table is mounted on the saddle. The table movements of the *plain* milling machine are

- *Vertical*—Raising the knee on the column;

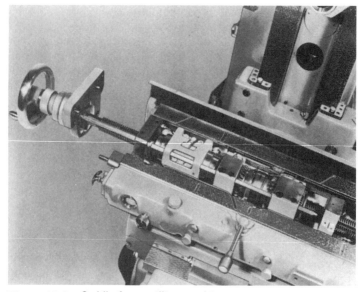

Figure 11-3 Saddle for a milling machine. *(Courtesy Cincinnati Milacron Co.)*

- *Transverse*—Sliding the saddle on the knee; and
- *Longitudinal*—Sliding the table on the saddle.

Micrometer dials graduated in thousandths of an inch permit accurate setting or positioning of the table. The table can also be swiveled horizontally on the saddle of the *universal* milling machine.

The milling machine table has T-slots running lengthwise on its top surface. T-bolts can be used to fasten either the work or a work-holding device to the table.

Types of Milling Machines

The knee-and-column type of milling machine is used more frequently in tool-and-die shops than any other kind of milling machine. The knee-and-column mill is classified as either *horizontal* or *vertical*, depending on the position of the spindle and table adjustments. The horizontal knee-and-column mill is divided into two classes: plain and universal.

Plain Milling Machine

The *plain horizontal milling machine* is very common in industry. The horizontal spindle projects at right angles from the column face. The spindle is hollow and tapered internally and permits the knee to be raised and lowered vertically on the accurately machined ways of the column. The table is mounted on machined ways on the saddle that rests on the knee. It can be moved either by hand or by power. The table feed (Figure 11-4) on plain milling can be in three directions—longitudinal (at right angles to the column face), transversal (crosswise), and vertical.

Figure 11-4 Feed unit for a milling machine. *(Courtesy Cincinnati Milacron Co.)*

Universal Milling Machines
The *universal milling machine* has the same table movements (longitudinal, crosswise, and vertical), and, in addition, the table can be set at an angle to the column face. The table is mounted on a swivel block that can be rotated about the center of the universal saddle. A modern universal milling machine is shown in Figure 11-5.

Vertical Milling Machine
The spindle is in a vertical position on the *vertical knee-and-column mills*. The spindle can be raised and lowered, and the machine is especially adapted for boring and profile work. Since the cutting tool is held in the spindle, an arbor is not needed on the vertical mill. The table of the vertical milling machine can be used on longitudinal, transverse, and vertical feed movements. A vertical milling machine is shown in Figure 11-6.

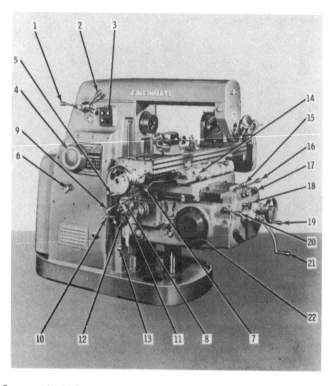

1. Overarm positioning lever	12. Rear feed-change knob
2. Overarm clamping lever	13. Rear vertical hand traverse
3. Rear push-button group	14. Automatic backlash eliminator knob
4. Spindle-speed dial	15. Rapid-traverse lever
5. Table traverse handwheel	16. Cross-feed lever
6. Spindle reverse lever	17. Front push-button
7. Rear table feed lever	18. Front feed-change knob
8. Rear cross-feed lever	19. Cross-traverse handwheel
9. Rear rapid-travers lever	20. Vertical-feed lever
10. Knee clamping lever	21. Vertical hand traverse crank
11. Rear cross hand traverse	22. Feed dial

Figure 11-5 Universal milling machine. *(Courtesy Cincinnati Milacron Co.)*

Special Types of Milling Machines

In addition to the knee-and-column milling machines, a number of specialized types of milling machines are used for certain kinds of work. *Thread mills* are used to mill both external and internal threads. *Spline mills* are used to cut keyways and slots.

In *bed-type* milling machines, the table cannot be raised or lowered. However, the spindle can be raised or lowered. The table moves in a horizontal position with longitudinal feed only. It does

Figure 11-6   Vertical milling machine. *(Courtesy Cincinnati Milacron Co.)*

not have a cross-feed movement. These are high-production machines.

The *duplex* milling machine has two opposed spindles. The spindles can be operated simultaneously, but each is separately driven and controlled. This type of machine has high-productive capacity, because two cuts can be taken at the same time on similar or different workpieces. A roughing cut with one cutter and a finishing cut with the other can be taken when the machine has two spindles.

Summary
The milling machine is a power-driven machine that cuts metal by using multitooth rotating cutters. The workpiece is fed to a rotary cutter instead of revolving as on a lathe or reciprocating as on a planer.

Many operations can be performed on the milling machine. The cutters remain sharp much longer than a single cutting tool, and the cutting action of the milling machine is continuous, as compared to the intermittent action of the shaper or planer. Large castings or forgings can be handled with ease on the milling machine.

The knee-and-column milling machine is used more frequently in tool-and-die shops than any other type. This machine is classified as either horizontal or vertical, depending on the position of the spindle and table adjustments. Knee-and-column milling machines are divided into two classes—plain and universal. There are also special types of milling machines (such as thread mills) used to mill both external and internal threads. A spline mill is used to cut keyways and slots. A duplex milling machine has two opposed spindles that can perform two separate operations at the same time.

Review Questions

1. What are the advantages in using a duplex milling machine?
2. Which milling machine is most common in tool-and-die shops?
3. What is the basic design of a milling machine?
4. Into what two classes are knee-and-column machines divided?
5. What operations can be performed on a milling machine?
6. Why is the spindle hollow throughout its entire length?
7. What are the table movements of the plain milling machine?
8. Which type of milling machine is common in industry?
9. What are thread mills used for?
10. What are bed-type milling machines used for?

Chapter 12

Milling Machine Operations

As discussed in Chapter 11, a milling machine cuts metal by means of a multitooth rotating cutter. The operator of a milling machine (or any other machine, for that matter) should be aware of certain safety precautions and operational guidelines. A milling machine is, of course, not a plaything; if proper precautions are taken, danger can be avoided.

Safety Precautions

First, the operator should understand thoroughly the construction and operating action of the machine before attempting its operation. This is important not only for the operation but also because it lessens the chances of the machine itself being damaged seriously.

The following suggestions by Brown and Sharpe Manufacturing Company should be carefully noted and remembered:

- Do not move an operating lever without knowing in advance the action that is going to take place.

- Never toy with the control levers or carelessly turn the handles of the milling machine while it is stopped.

- Do not lean against, or rest the hands on, a moving table; if it is necessary to touch a moving member, make sure of the direction in which it is moving.

- Do not attempt to cut without being sure that the work is held securely in the vise or fixture and that the holding member is fastened to the machine table.

- Remove chips with a brush or other suitable means—never with the fingers or hands.

- Before operating a milling machine, study it thoroughly. Then, if an emergency should arise, the machine can be stopped immediately.

- Stay clear of the milling cutters; do not touch a cutter, even while it is not moving, unless there is good reason to do so, and then be careful.

The above suggestions, in general, can be applied to any other machine in the machine shop.

Preliminary Operations

The various duties of the milling machine operator can be classified as preliminary operations and machining operations. Several preliminary operations are necessary before the machine can be started.

Cleaning

Cleaning of the milling machine cannot be emphasized too strongly. The machine should be cleaned both before and after using. Accuracy and durability of the machine depend on its being kept clean; this applies to all machine tools.

If a milling machine is cleaned as soon as possible after use, it can be kept clean more easily than if chips and coolant are allowed to accumulate for long periods of time. A clean machine is usually in better mechanical condition than one in which dirt has been allowed to accumulate. This is because chips that build up on a dirty machine can clog oil channels, mar the bearing surfaces of the column or knee, and interfere with accurate adjustments of the fixtures and work. Oil holes that have become clogged with gummy oil can be flushed with gasoline; this will not injure the bearings.

Oiling the Machine

Apply a few drops of oil to each bearing at frequent intervals, rather than a flood of lubricant at long or irregular intervals. Most bearings can hold only a few drops, and an excess of oil runs out and is lost. It is always safest to purchase a lubricant that is reliable, rather than experiment with less expensive machinery oils. It is less expensive to use a good oil than to risk damage to bearings from overheating or scoring.

The more recently designed milling machines are equipped with automatic lubricating systems that ensure a constant supply of clean lubricants at important points. This relieves the machine operator of much of the oiling problem, but the oil level in machine reservoirs should be checked at regular intervals.

Many machines equipped with automatic lubrication include oil filters in the system. All oil must pass through the filters each time it is circulated. The filters should be checked regularly to determine whether they are functioning properly. When the filter becomes clogged, the bearings can become dry, even though the oil is bypassed around the filter.

Mounting the Workpiece

Methods of mounting the work on the milling machine are similar to those for the shaper, planer, and other machines with worktables. However, work should be mounted more securely on the milling machine tables because the greater thrust produced by the cutter (especially the wider cutters) can damage either the work or the machine. Various methods of mounting work on the milling machine for machining are on the table, in a chuck, on an arbor, between centers, and on fixtures.

Clamping devices (such as T-bolts, toe dogs, pins, and screw pins) are similar to those used to hold planer workpieces (Figure 12-1).

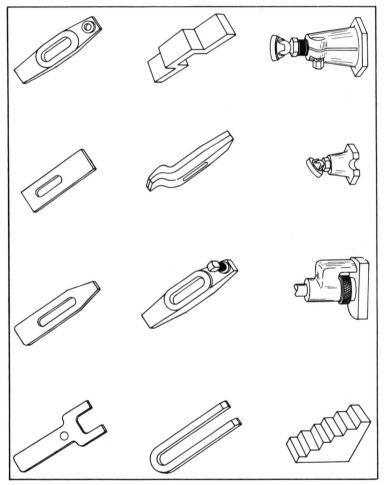

Figure 12-1   Types of clamps, step blocks, and jacks commonly used in milling machine setups. *(Courtesy Cincinnati Milacron Co.)*

The application of these devices is the same as for other machines, except that even more attention is required for supporting and anchoring the work to prevent springing. Stop pins should be used wherever possible to prevent movement of the work.

Much of the work that is performed on milling machines is mounted on fixtures. The table, fixtures, and all devices involved in mounting the work should be absolutely free from chips, dust, and so on. Fixtures can be built to facilitate handling and to increase production if a large enough quantity of pieces warrants the expense.

For single-piece production or small-lot production, the work must be held by individual setup because it is impractical to construct special fixtures. The work must be positioned and held securely. The clamps must be carefully arranged, so that they will not be struck by the cutter or damage other parts of the machine.

The milling machine vise is the simplest holding arrangement. A special base for bolting to the table and a swiveling body permit adjustment of the vise to any angle in a horizontal plane. A scale (graduated in degrees) on the base of the vise is accurate enough for most work. However, if accurate angles are required, the setting should be checked with a protractor. The toolmaker's universal vise can be used for work involving compound angles (Figure 12-2). It is designed for adjustment in both the vertical and horizontal planes.

Figure 12-2   The toolmaker's universal vise can be used for general toolroom work. It swivels 90° in the vertical plane and 360° in the horizontal plane. *(Courtesy Cincinnati Milacron Co.)*

The standard vise is opened and closed by a crank. Hammering the handle of the vise to obtain a sure grip is often frowned upon because it can crack or break the screw or other parts, resulting in a serious accident when pressure is applied. However, many practical machinists insist upon the practice, maintaining that the work must be held securely. If the practice is followed, use a lead or Babbitt hammer for a single sharp blow, after tightening by hand as much as possible. A machinist's hammer should never be used on the handle of the vise.

The base of the vise should be cleaned carefully before installing it on the machine. Then, a dial indicator can be used to check for accurate alignment (Figure 12-3). Clamp the dial indicator onto the machine arbor; then, move the table so that the indicator rests on the fixed (not the adjustable) jaw. Set the indicator dial to zero, and by means of the manual feed, traverse the jaw past the indicator. Adjust the vise until the indicator shows no variation between the two ends of the jaws.

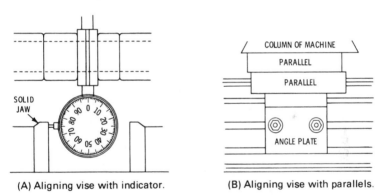

(A) Aligning vise with indicator.   (B) Aligning vise with parallels.

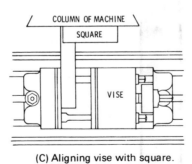

(C) Aligning vise with square.

Figure 12-3 Methods of aligning the vise jaws with the columns or spindle. The dial indicator method is the most accurate. *(Courtesy Cincinnati Milacron Co.)*

Another method of aligning the vise is to place the stock of a machinist's square against the face of the machine column and bring the blade into contact with the fixed jaw—with tissue paper feelers between the blade and the jaw. If an extra pull is required to remove all feelers, the vise is aligned (see Figure 12-3).

Selecting the Cutter
Failure to obtain satisfactory results on a job can often be attributed to improper selection of the milling cutter. Even if the proper cutter is used, working conditions can be unfavorable to proper performance of the cutter. Either the operator or another person in the milling department should be proficient in the use and care of milling cutters and capable of determining the correct feeds and speeds for their operation. Following is a brief summary of the general application of various milling cutters:

- *Double-angle cutter*—Used for fluting taps, reamers, and so on.
- *End mill*—Suited for surface milling, profiling, slotting, and so on.
- *Fly cutter*—Can be used to mill intricate shapes that do not warrant the expense of formed cutters.
- *Formed cutters*—Used to cut curved or irregular surfaces; also for fluting taps, reamers, twist drills, and so on.
- *Half side cutter*—Used for face milling, or in pairs for "straddle milling."
- *Helical cutter*—Especially adapted for thin work or intermittent cuts where the amount of slack to be removed varies. It is not a general-purpose cutter. This cutter is erroneously called a "spiral" cutter.
- *Inserted-tooth cutter*—Characterized by long life; it is used for heavy-duty face and side milling work.
- *Interlocking side cutters*—Designed to maintain an exact width for milling slots.
- *Nicked cutters*—Suited for making deep cuts, as in roughing work.
- *Plain cutters*—Used to mill a flat surface parallel to its axis.
- *Side cutter*—Suitable for slotting operations.
- *Single-angle cutter*—Chiefly used to cut milling machine cutter teeth.

- *Slitting cutter*—Used to cut deep slots and to cut off stock.
- *Slotting cutter*—Used to mill shallow slots, such as screw-head slots.

In addition to this summary of cutter applications, the final selection of the proper milling cutter can be influenced by other factors, such as the following:

- Milling cutters with comparatively few widely spaced teeth can remove more metal in a given time (for certain kinds of milling work) than cutters with a comparatively large number of teeth, without stressing the cutter or overloading the machine.
- Fine-tooth cutters are inferior to coarse-tooth cutters, so far as their relative tendency to "chatter" is concerned.
- The free cutting action of coarse-teeth cutters is primarily because less cutting action is required to remove a given amount of metal; each tooth takes a larger, deeper chip.
- Wide spaces between the teeth allow the cutting edges to be well backed up, which is not always possible with teeth that are closely spaced.
- Moderate rake angles reduce power consumption and are desirable on cutters that are used on mild steel. Large rake angles are undesirable because of the tendency to chatter.
- The helical angle has little effect on power consumption. However, a large helical angle is desirable because it requires fewer cutter teeth, gives smoother cutter action, and reduces the tendency to chatter.
- With few exceptions, coarse-tooth cutters are superior to fine-tooth cutters for production work.

Sharpening Milling Cutters
Do not permit a cutter to become dull. Sharpen the cutter often. Experience has shown that a dull cutter wears more rapidly than a sharp one and does not last as long as a cutter kept in proper condition.

The clearance or relief of a milling cutter is the amount of material removed from the top of the teeth, back of the cutting edge. This permits a tooth to clear the stock after the cutting edge has done its work. On formed cutters, it is not necessary to consider clearance. This is because the teeth are formed so that the clearance angle remains the same when the face of a tooth is ground. Rake and clearance angles are illustrated in the profile of a milling cutter shown in Figure 12-4. The narrow space or land *A* back of the cutting edge is about ⅟₃₂ inch in width and is ground to a clearance angle *B*. This clearance angle, together with the rake angle *C* on the front of the tooth, varies with the material being cut. Changing either the clearance angle or the rake angle (or both) only slightly can make a vast difference in the operation of the milling cutter on the workpiece.

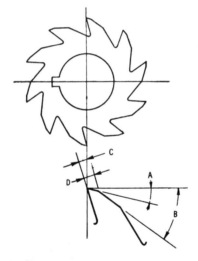

Figure 12-4 Profile of a milling cutter showing (A) clearance angles, (B) secondary clearance, (C) rake or undercut, and (D) land.

The clearance angles of milling cutters should always be checked. The fact that a milling cutter has a keen cutting edge does not necessarily mean that it will cut satisfactorily. As a general rule, the clearance angle for plain milling cutters more than 3 inches in diameter should be 4°. The clearance angle for plain cutters less than 3 inches in diameter should be 6°. The Cincinnati Milacron Company recommends the following clearance angles:

- *Aluminum*—10°
- *Bronze, cast*—10–15°
- *Cast iron*
  - Fast feeds—7°
  - Moderate feeds—6–7°
- *Copper*—7–10°

- *Steel*
  - Castings—6–7°
  - Hard (tool steel)—4–5°
  - Low-carbon—5–7°

The clearance angle for helical mills depends on the material being milled, depth of cut, and so on. Generally, a 5° to 7° clearance angle for cast iron and a 3° to 4° clearance angle for machine steel are satisfactory.

Details of a correctly sharpened side mill for milling cast iron are shown in Figure 12-5. The illustration shows a ³⁄₆₄-inch land that is ground at an angle of 6°, which is the clearance angle. Immediately behind the land, the tooth of the cutter is ground at an angle of about 12°. This angle should be no larger than is necessary to prevent the heel of the cutter dragging on the work. The side teeth should be ground in exactly the same manner—a ³⁄₆₄-inch land, a 6° clearance angle, and backed off at a 12° angle. If the side mill were used on steel, the proper clearance angle would be 4°, rather than 6°. If this kind of cutter shows a tendency to chatter, the clearance angle on the side teeth should be reduced to as small as 1°. Conditions can be improved even more if the sides are slightly hollow ground, that is, if the face of the cutter is ground thinner at the inner end of the side teeth (*A* in Figure 12-5) than at the outer end *B*. Fixed rules cannot be given for the clearance angles on cutters, as this depends on the material being machined, the depth of the cut, the style of the cutter, and so on.

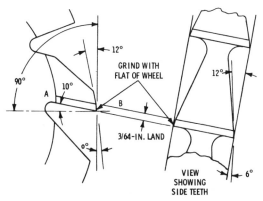

Figure 12-5 Details of a correctly sharpened milling cutter.

Mounting the Cutter and Arbor

To obtain rigidity in the setup, select a cutter that is large enough to prevent using an arbor of large diameter. This is necessary to obtain best results in heavy-duty milling. Place the cutter (or cutters) on the arbor as near the end of the spindle as the work permits to prevent springing the arbor (Figure 12-6).

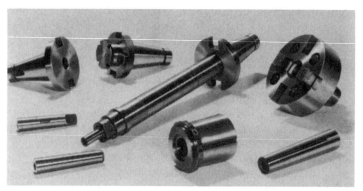

Figure 12-6 Typical milling machine arbors. *(Courtesy Cincinnati Milacron Co.)*

The support arm, which carries the outer bearing and the intermediate bearings, should always be used to provide proper support for the arbor (Figure 12-7). The arbor support arm carrying the outer bearing should be in place before tightening the arbor nut to avoid danger of bending the arbor.

Milling cutters preferably should have a key drive rather than a friction drive. If a key drive is used, it is not necessary to hammer the arbor wrench to tighten the nut; it can be tightened sufficiently by hand. Various cutter mounting and arbor supports are shown in Figure 12-8.

Short arbors (see Figure 12-8A) are provided with a pilot bearing at the end of the arbor. This bearing fits a split bronze bushing X in the arbor support.

Some medium-length arbors (see Figure 12-8B), in addition to an end pilot bearing X, have an arbor bearing collar to fit the intermediate support Y. This support should be placed as close to the milling cutter as is practical; the cutter should be located as close to the shoulder of the arbor as conditions permit. Another style of medium-length arbor (see Figure 12-8C) does not have the pilot bearing for the bronze bushings at the end of the arbor, but it is provided with a bearing collar for placing the arbor support near the cutter.

Figure 12-7 An overarm provides rigid support for the arbor and cutters, maintains accurate alignment, and dampens vibrations set up by cutter action. *(Courtesy Cincinnati Milacron Co.)*

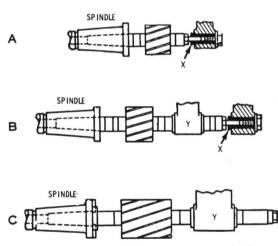

Figure 12-8 Various cutter mountings and arbor supports for a short arbor (A) and medium-length arbors (B and C).

Long arbors (Figure 12-9A) should have two support bearings whenever possible. One support Y should be placed between cutters that are spaced at a distance on the arbor, and the other support Z (to which the braces are fastened) should be as near the outside cutter as conditions permit.

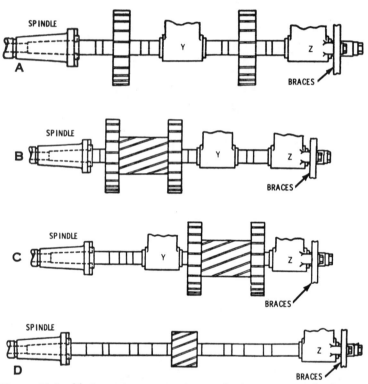

Figure 12-9   Various cutter mountings and arbor supports for long arbors. Correct methods of mounting are shown in (A), (B), and (C); an incorrect method of mounting is shown in (D).

If the width of the table does not permit bringing the outer support Z (to which the braces are fastened) near the cutters, the intermediate support Y can be placed near the outer cutter, between it and the outer support (see Figure 12-9B).

In some instances, the nature of the work requires the cutters to be placed near the outer end of the arbor (see Figure 12-9C). Then, the intermediate support *Y* should be placed between the inner cutter and the spindle.

An incorrect method of mounting a milling center on a long arbor is shown in Figure 12-9D. This kind of setup should never be used, because it cannot possibly produce results that are satisfactory.

Direction of Cutter Rotation

The direction of cutter rotation in relation to direction of feed is highly important. Many cutters can be reversed on the arbor, so it is important to know whether the spindle is to rotate clockwise or counterclockwise. In this connection it is also important to know the direction of feed. The direction of feed is opposite the direction of cutter rotation in *conventional milling* (Figure 12-10). This can be called "up-cutting," which means that the cutting edge starts at the bottom of the cut and removes a progressively larger chip as the feed progresses. Formerly, all milling was done in this manner, but it has a disadvantage in that a certain amount of rubbing occurs before the tooth takes hold, and a series of small scallops shows on the finished surface.

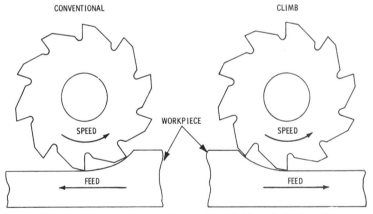

Figure 12-10   Conventional milling action (left) and climb milling action (right).

The direction of feed is in the same direction as the direction of cutter rotation in *climb milling* (see Figure 12-10). This action is also known as "down milling" or "hook milling." Fast cutting rates and a better finish are obtained in climb milling. The pressure of the cut tends to hold the workpiece downward on the table. However, climb milling cannot be used on older machines unless they have a backlash eliminator on the feed screw nut—except where the work must be climb-milled and light cuts can be taken. If climb milling is attempted on an unsuitable machine, the work can be dragged under the cutter, which will jam and break, and the machine can be seriously damaged. Most modern milling machines are designed for climb milling.

Locating the Cutter

For most milling operations, the position of the cutter in relation to the work is not of great importance. Usually the worktable can be moved up or down and forward or backward until the cutter barely touches the work. If it is necessary to position the cutter accurately in relation to a flat surface, a square or straightedge can be used to align the cutter with the surface. Then, the work can be moved the necessary amount by means of the manual controls.

If a round surface is involved (such as milling a keyway in a shaft), a different positioning technique is required (Figure 12-11). If the front side of the cutter can be brought tangent to the shaft, move the table by hand until a tissue paper feeler can barely be removed from between the work and the cutter. Set the cross-feed dial to zero. Then, lower the worktable and move the workpiece toward the cutter a distance equal to one-half the cutter thickness (see Figure 12-11A). This distance can be measured by carefully counting the graduations on the dial. Even the paper thickness can be allowed for, if extreme accuracy is desired.

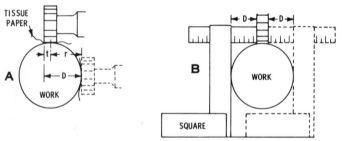

Figure 12-11 Two methods of locating the cutter in relation to the center of a round workpiece: (A) tissue paper method, and (B) use of square and scale.

If the cutter cannot be brought tangent to the workpiece, a square and scale can be used (see Figure 12-11B). To set for depth of cut, raise the worktable until a paper feeler can barely be removed from between the cutter and the top of the workpiece, and set the vertical-feed dial to zero.

Speeds and Feeds

Speed and feed rates are governed by several variable factors: material, cutter, width and depth of cut, required surface finish, machine rigidity and setup, power and speed available, and cutting fluid. Even though suggested rate tables are provided here, individual experience and judgment are extremely valuable in selecting correct milling speeds and feeds. The lower figure in the table for a given material should always be used until sufficient practical experience has been gained to change to the highest figure. Speed can be increased until either excessive cutter wear or chatter indicates that the practical limit has been exceeded.

Speeds

Unlike lathe tools, the cutting action for each tooth of a milling cutter is intermittent; there are relatively long cooling intervals, with the cooling intervals for large-diameter cutters being longer, of course, than for smaller cutters. The introduction of high-speed steel and the newer, faster-cutting alloys for construction of the cutters has considerably increased cutter speeds, as compared with the cutting speeds for the carbon steel cutters that were used formerly. Thus, the material used in construction of the cutters has an important bearing on permissible speeds (see Table 12-1).

Table 12-1   Cutting Speeds (Surface Feet per Minute)

Material	High-Speed Steel		Carbide-Tipped		Coolant
	Rough	Finish	Rough	Finish	
Cast iron	50–60	80–110	180–200	350–400	Dry
Semisteel	40–50	65–90	140–160	250–300	Dry
Malleable iron	80–100	110–130	250–300	400–500	Soluble, sulfurized, or mineral oil
Cast steel	45–50	70–90	150–180	200–250	Soluble, sulfurized, mineral, or mineral lard oil
Copper	100–150	150–200	600	1000	Soluble, sulfurized, or mineral lard oil
Brass	200–300	200–300	600–1000	600–1000	Dry

(continued)

<table>
<tr><td colspan="7" align="center">Table 12-1 (continued)</td></tr>
</table>

Material	High-Speed Steel		Carbide-Tipped		
	Rough	Finish	Rough	Finish	Coolant
Bronze	100–150	150–180	600	1000	Soluble, sulfurized, or mineral lard oil
Aluminum	400	700			Soluble or sulfurized oil, mineral oil, and kerosene
Magnesium	600–800	1000–1500	1000–1500	1000–5000	Dry, kerosene, mineral lard oil
SAE steels					
1020 (coarse feed)	60–80	60–80	300	300	Soluble, sulfurized, mineral, or mineral lard oil
1020 (fine feed)	100–120	100–120	450	450	Soluble, sulfurized, mineral, or mineral lard oil
1035	75–90	90–120	250	250	Soluble, sulfurized, mineral, or mineral lard oil
X-1315	175–200	175–200	400–500	400–500	Soluble, sulfurized, mineral, or mineral lard oil
1050	60–80	100	200	200	Soluble, sulfurized, mineral, or mineral lard oil
2315	90–110	90–110	300	300	Soluble, sulfurized, mineral, or mineral lard oil
3150	50–60	70–90	200	200	Soluble, sulfurized, mineral, or mineral lard oil
4340	40–50	60–70	200	200	Sulfurized and mineral oils
Stainless steel	100–120	100–120	240–300	240–300	Sulfurized and mineral oils

Courtesy Cincinnati Milacron Co.

The amount of material to be removed per minute, and the relationship between depth of cut and feed, influences cutter speed. For example, a cut ⅛ inch in depth and ⅛-inch feed per revolution can be taken at a higher speed than a cut ¼ inch in depth and ¹⁄₁₆-inch feed per revolution, although the amount of material removed per minute would be the same in each operation. Cutter speed also can depend on the rigidity of the machine and the fixture in which the workpiece is held.

On some jobs, disregarding expense, the cutter can be run at excessive speed if it is reground frequently, or it can be run at a very slow speed and reground at longer intervals. Accordingly, speeds can be determined for a given cutter at which its greatest efficiency is achieved.

Cutter speeds are always given in surface feet per minute (sfpm), that is, the speed at which the circumference of the cutter passes over the work. Thus, the surface speed of a 6-inch cutter is twice as great as that of a 3-inch cutter for a given number of spindle revolutions. As spindle speed of a milling machine can be given only in revolutions per minute (rpm), it is necessary to translate sfpm into rpm for making speed adjustments on the machine. If a table of speeds is not available, spindle speed can be calculated by the following formula:

$$rpm = \frac{sfpm \times 12}{diameter}$$

Likewise, sfpm can be calculated by the following formula:

$$sfpm = \frac{diameter \times rpm}{12}$$

Feeds

The feed rate is the rate at which the work advances past the cutter and is commonly given in inches per minute (in./min.). Generally, the rule in production work is to use all the feed that the machine and the work can stand. However, it is a problem to determine where to start the feed. Table 12-2 and Table 12-3 provide the suggested feed per tooth for high-speed steel and sintered carbide-tipped milling cutters, respectively. Note that the feeds are given in thousandths of an inch per tooth for the various cutters. Multiply the feed per tooth by the number of teeth, and multiply that product by the rpm to determine the feed rate in in./min., as follows:

$$in./min. = feeds\ per\ tooth \times number\ of\ teeth \times rpm$$

Table 12-2 Suggested Feed per Tooth for High-Speed Steel Milling Cutters

Material	Face Mills	Helical Mills	Slotting and Side Mills	End Mills	Form-Relieved Cutters	Circular Saws
Plastics	0.013	0.010	0.008	0.007	0.004	0.003
Magnesium and alloys	0.022	0.018	0.013	0.011	0.007	0.005
Aluminum and alloys	0.022	0.018	0.013	0.011	0.007	0.005
Free-cutting brasses and bronzes	0.022	0.018	0.013	0.011	0.007	0.005
Medium brasses and bronzes	0.014	0.011	0.008	0.007	0.004	0.003
Hard brasses and bronzes	0.009	0.007	0.006	0.005	0.003	0.002
Copper	0.012	0.010	0.007	0.006	0.004	0.003
Cast iron, soft (150–180 BH)	0.016	0.013	0.009	0.008	0.005	0.004
Cast iron, medium (180–220 BH)	0.013	0.010	0.007	0.007	0.004	0.003
Cast iron, hard (220–300 BH)	0.011	0.008	0.006	0.006	0.003	0.003
Malleable iron	0.012	0.010	0.007	0.006	0.004	0.003
Cast steel	0.012	0.010	0.007	0.006	0.004	0.003
Low-carbon steel, free machining	0.012	0.010	0.007	0.006	0.004	0.003
Low-carbon steel	0.010	0.008	0.006	0.005	0.003	0.003
Medium-carbon steel	0.010	0.008	0.006	0.005	0.003	0.003
Alloy steel, annealed (180–220 BH)	0.008	0.007	0.005	0.004	0.003	0.002
Alloy steel, tough (220–300 BH)	0.006	0.005	0.004	0.003	0.002	0.002
Alloy steel, hard (300–400 BH)	0.004	0.003	0.003	0.002	0.002	0.001

Table 12-2   *(continued)*

Material	Face Mills	Helical Mills	Slotting and Side Mills	End Mills	Form-Relieved Cutters	Circular Saws
Stainless steels, free machining	0.010	0.008	0.006	0.005	0.003	0.002
Stainless steels	0.006	0.005	0.004	0.003	0.002	0.002
Monel metals	0.008	0.007	0.005	0.004	0.003	0.002

Courtesy Cincinnati Milacron Co.

Table 12-3   Suggested Feed per Tooth for Sintered Carbide-Tipped Cutters

Material	Face Mills	Helical Mills	Slotting and Side Mills	End Mills	Form-Relieved Cutters	Circular Saws
Plastics	0.015	0.012	0.009	0.007	0.005	0.004
Magnesium and alloys	0.020	0.016	0.012	0.010	0.006	0.005
Aluminum and alloys	0.020	0.016	0.012	0.010	0.006	0.005
Free-cutting brasses and bronzes	0.020	0.016	0.012	0.010	0.006	0.005
Medium brasses and bronzes	0.012	0.010	0.007	0.006	0.004	0.003
Hard brasses and bronzes	0.010	0.008	0.006	0.005	0.003	0.003
Copper	0.012	0.009	0.007	0.006	0.004	0.003
Cast iron, soft (150–180 BH)	0.020	0.016	0.012	0.010	0.006	0.005
Cast iron, medium (180–220 BH)	0.016	0.013	0.010	0.008	0.005	0.004
Cast iron, hard (220–300 BH)	0.012	0.010	0.007	0.006	0.004	0.003
Malleable iron	0.014	0.011	0.008	0.007	0.004	0.004
Cast steel	0.014	0.011	0.008	0.007	0.005	0.004
Low-carbon steel, free machining	0.016	0.013	0.009	0.008	0.005	0.004

(continued)

Table 12-3   (continued)

Material	Face Mills	Helical Mills	Slotting and Side Mills	End Mills	Form-Relieved Cutters	Circular Saws
Low-carbon steel	0.014	0.011	0.008	0.007	0.004	0.004
Medium-carbon steel	0.014	0.011	0.008	0.007	0.004	0.004
Alloy steel, annealed (180–220 BH)	0.014	0.011	0.008	0.007	0.004	0.004
Alloy steel, tough (220–300 BH)	0.012	0.010	0.007	0.006	0.004	0.003
Alloy steel, hard (300–400 BH)	0.010	0.008	0.006	0.005	0.003	0.003
Stainless steels, free machining	0.014	0.011	0.008	0.007	0.004	0.004
Stainless steels	0.010	0.008	0.006	0.005	0.003	0.003
Monel metals	0.010	0.008	0.006	0.005	0.003	0.003

Courtesy Cincinnati Milacron Co.

In actual practice, it is better to start the feed rate at a lower figure than that indicated in the table and to work upward gradually until the most efficient removal rates are reached. Too high a feed rate is indicated by excessive cutter wear. A cutter can be spoiled by too fine a feed, as well as by too heavy a feed. A rubbing action, rather than a cutting action, can dull the cutting edge, and excessive heat can be generated. This can be true also in relation to depth of cut. The first cut on castings and rough forgings should be made well below the surface skin. Milling cuts less than 0.015 inch should be avoided. To obtain a good finish, take a roughing cut followed by a finishing cut, with a higher speed and lighter feed for the finishing cut.

In general, feeds can be increased as speeds are reduced. Therefore, the feed can be increased for abrasive, sandy, or scaly material, and for heavy cuts in heavy work. Feed should be increased if cutter wear is excessive or if there is chatter.

Feeds should be decreased for a better finish, when taking deep slotting cuts, or when the work cannot be held rigidly. If the cutter begins to produce long, continuous chips, the feed should be decreased.

A good commercial finish can be obtained by using a feed rate that removes a chip of 0.030 inch to 0.050 inch per revolution of the cutter. Finer finishes, of course, necessitate finer feeds, but a feed of 0.015 inch to 0.020 inch per revolution usually produces an excellent finish.

Coolants

Various liquids and compressed air are used to carry off some of the heat generated in machining of metals. The kind of coolant depends on the material being machined.

Cast iron machined at usual speed and feed does not overheat the milling cutters. A liquid coolant should not be used with cast iron because the chips mix with the liquid, resulting in a sticky mass that clogs the cutter teeth and is difficult to remove.

Compressed air can keep the cutter cool and free from chips. The chief disadvantage of compressed air is that too much pressure can scatter the chips and dust, which makes the machine untidy and causes trouble.

Brass can be milled dry. Steel generates considerable heat in machining. Therefore, an abundance of coolant should be used. Lard oil makes an excellent coolant. The use of a coolant makes both light and heavy cuts possible at much higher speeds, thereby permitting a smaller cut per tooth, which reduces the stresses on the workpiece and the arbor.

The large volume of lubricant carries away most of the chips, which aids in reducing cleaning on some jigs and fixtures. A lubricant is usually applied to the cutter through piping that has a swivel joint, so that the liquid can be directed to the cutter. See Table 12-1 for recommended coolants to be used with various materials.

Taking the Milling Cut

A quick check should be taken to make sure that everything is in order before you actually begin to cut metal. Make certain that the work and fixture will clear all parts of the machine and that the cutter will not strike any part of the fixture or holding device. The machine should be well oiled, and there should be an adequate supply of cutting fluid in the reservoir. All table movements that will not be used in the operation should be locked, and all those movements that will be used should be unlocked. The table dogs should be set to throw off the feed after the required length of cut. The dogs should be set to stop the rapid traverse movement before the work reaches the cutter, if rapid traverse is to be used. The starting lever and table control levers should all be in neutral. The spindle speed control should be set to the required speed, and the feed rate should be set at one-half to one-quarter of the feed rate selected for cutting.

After starting the main motor and cutting fluid pump, push over the starting lever and make certain that the cutter is rotating in the correct direction. If it is rotating in the wrong direction, stop the spindle, engage the reverse lever, and restart the spindle. The work should be positioned for a light trial cut before going to full depth. Push the table control lever in the direction the table is to move, and take the trial cut for about ½ inch in length. Then, stop the machine, reverse the table, and examine the cut. If the trial cut is satisfactory, increase the feed rate to the full rate selected. Take a full-depth cut over the entire workpiece surface, making certain that plenty of cutting fluid is pouring over the point where the cutter engages the metal. Unless there is a very good reason, do not stop the cutter until it has passed entirely over the work, or else a depression or scallop will be left on the surface. This is particularly important on finishing cuts.

The work should not be reversed under the cutter without first backing it away slightly. To do so can cause chatter and can cause a jam that can break the cutter and spoil the work. All feed screws have

Figure 12-12 The universal dividing head permits indexing through any desired number of divisions. *(Courtesy Cincinnati Milacron Co.)*

backlash, so a slight backward movement on the handle will not move the table the same amount. Note the starting point, and then give the handle at least a half turn. When the handle is returned to the starting point, the work will be returned to exactly the same place; then the handle can be advanced the required amount for the next cut.

Attachments and Accessories

A wide variety of standard attachments and accessories can increase the overall usefulness of the standard milling machine. One of the most useful of these is the *universal dividing head* (Figure 12-12). This device provides a means of holding the work in a chuck, between the centers, or in a collet, and revolving it (indexing) through a desired number of equal divisions. The unit is mounted on the machine table, and the head can be tilted to any angle. The work can be rotated through a full 360° by a worm or bevel gears.

Figure 12-13   Micrometer table attachment for use on a milling machine. *(Courtesy Cincinnati Milacron Co.)*

A series of index plates is available for dividing the circle into any desired number of parts. For example, to divide into seven parts, select a circle of holes in the index plate that is divisible by 7, such as 28. As 40 revolutions of the crank are necessary for one revolution of the work on the spindle, divide 40 by the number of divisions (40 ÷ 7) to give $5\frac{5}{7}$. Therefore, to obtain seven equal divisions, it is necessary to turn the crank five complete turns and a $\frac{5}{7}$ turn for each division. As the circle has 28 holes, use $\frac{5}{7}$ of 28, or 20 holes. So, after each cut, turn the crank five complete turns, plus 20 spaces on the 28-hole circle.

Figure 12-14 Milling a dovetail with the vertical milling attachment and swivel vise. *(Courtesy Cincinnati Milacron Co.)*

A *micrometer table attachment* for jig and die work is shown in Figure 12-13. This attachment enables the operator to perform boring and recessing operations.

Vertical milling attachments are designed so that vertical milling can be performed on a horizontal machine (Figure 12-14). These attachments can be used for drilling, boring, cutting T-slots, and other such work.

Summary

The operator of a milling machine should be aware of certain safety precautions. If proper precautions are taken, danger can be avoided and safe, profitable operations can be assured. Never move an operating lever without knowing in advance the action that is going to take place. Stay clear of the milling cutter; do not touch a cutter unless there is a good reason to do so. These are just a few precautions that should be observed. Keeping the machine clean and oiled cannot be emphasized too strongly. A clean and oiled machine is generally in better mechanical condition than one on which dirt has been allowed to accumulate.

Workpieces must be mounted properly and securely. Clamping devices (such as T-bolts, toe dogs, and screw pins) are similar to those used to hold planer workpieces. Always keep cutters clean and sharp. Dull cutters wear more rapidly than sharp ones and do not last as long as cutters kept in proper condition.

Proper speed and feed rates are governed by several factors. These factors are material, cutter width, depth of cut, surface finish, power and speed available, and cutting fluid. The amount of material to be removed per minute is in direct relation to the depth of cut and cutter speed. The feed rate is the rate at which the work advances past the cutter, which is commonly referred to as inches per minute.

Various liquids and compressed air are used to carry off some of the heat generated in machining of metals. The kind of coolant depends on the material being machined. Compressed air can keep the cutter cool and free from chips. The chief disadvantage of compressed air is that too much pressure can scatter the chips and dust, which makes the machine untidy and causes trouble.

A quick check should be taken to make sure that everything is in order before you actually begin to cut metal.

A wide variety of standard attachments and accessories can increase the overall usefulness of the standard milling machine. One of the most useful of these is the universal dividing head.

Review Questions

1. What should the operator of a milling machine do to make the machine a safe piece of equipment?

2. Why should the machine be kept clean and properly oiled?

3. What is meant by clearance angle?

4. List some of the mounting methods and devices used with a milling machine.

5. Why is selecting the proper milling machine cutter so important?

6. Why is sharpening a milling machine cutter so important?

7. What is the purpose of the overarm on a milling machine?

8. What is an arbor?

9. Why is the direction of cutter rotation in relation to the direction of feed so important?

10. What is climb milling?

Chapter 13

Milling Machine Dividing Heads

The *dividing head* (also called an *index head*) is a device that can be used to rotate a piece of work through given angles—usually equal divisions of a circle. A dividing head, in combination with the longitudinal feeding movement of the table, is used to impart a rotary motion to a workpiece for helical milling action, such as in milling the helical flutes of cutters.

Dividing heads are used in milling operations whenever it is necessary to divide a circle into two or more parts (Figure 13-1). This includes spacings for gear teeth or milling cutter teeth, boring holes in jigs, and other kinds of work that require precision spacing. When a tailstock is used with the dividing head to support the workpiece between centers, the combination setup is called an *index center*. When it is in use, the dividing head is bolted to the table of the milling machine and becomes part of the equipment of the machine.

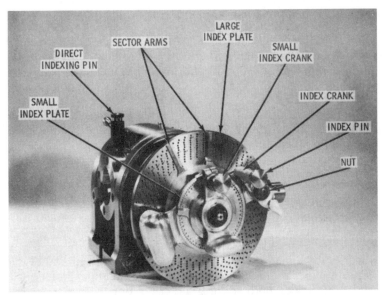

Figure 13-1   Universal wide-range divider for milling machine.
(Courtesy Cincinnati Milacron Co.)

Classification

The types of dividing heads used on milling machines are plain, universal, and helical. Dividing heads can also be classified as to size (for example, the Cincinnati Milacron Company dividing head is available in three different sizes).

Plain Dividing Head

The basic parts of a *plain dividing head* are shown in Figure 13-2 and Figure 13-3. In a plain dividing head, the spindle rotates about a horizontal axis.

The principal parts of the dividing head are spindle, worm wheel, worm, index plate, index pin, and sector.

The *worm wheel* is keyed to the *spindle* so that the spindle turns with the wheel. The worm wheel has 40 teeth on most dividing heads.

The *worm* meshes with the worm wheel. The ratio of the two gears is 40:1. Therefore, 40 revolutions of the worm are required to turn the worm wheel one complete revolution.

The *index plate* is one of a set that is provided with the dividing head. The index plate can be removed easily and another plate substituted as necessary for the desired spacing. Each index plate consists of several circular rows of holes—each circular row having a different number of holes.

The *index pin* is located on the end of the crank, which is attached to the worm shaft. The crank is used to rotate the spindle through the worm gearing. The arm length of the crank is adjustable so that the index pin can drop into any hole in any circular row of holes.

The *sector* consists of two radial arms constructed in such a manner that the angle between them can be changed and locked by a clamp screw in any included angular position. The sector can be used to save time and reduces the possibility of an error in counting the number of holes for each movement of the index pin.

In actual operation, the index crank should be adjusted for the correct circular row of holes. The index pin is dropped into the correct hole in the row. Then, rotate the sector arm A (see Figure 13-2) against the left-hand side of the index pin. Next, move the sector arm B in the same direction that the index pin is to turn until the correct number of holes is counted between the pin and the sector arm B. Now lock both sector arms in position. *Never* count the hole in which the index pin has been inserted; this hole is the "zero" hole.

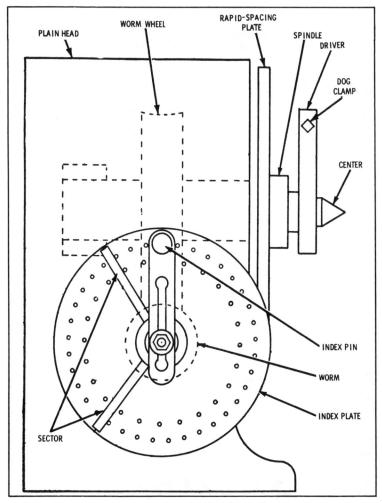

Figure 13-2 Diagram showing basic parts of a plain dividing head (side view).

Universal Dividing Head

The *universal dividing head* differs from the plain dividing head in that the spindle can be tilted (that is, swiveled to any angular position in a vertical plane, within the angular range provided). As shown in Figure 13-4, the casting that carries the spindle is

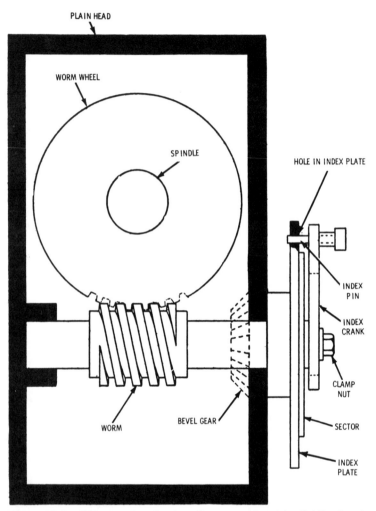

Figure 13-3 Diagram showing interior parts of a plain dividing head (cross-sectional view).

mounted in a circular guide, forming part of the swivel or universal head. The spindle can be tilted to any angular position *AB* by rotating the spindle casting, which can be clamped in any angular position by tightening the angular clamp bolt. The circular guide is graduated in degrees and fractional degrees.

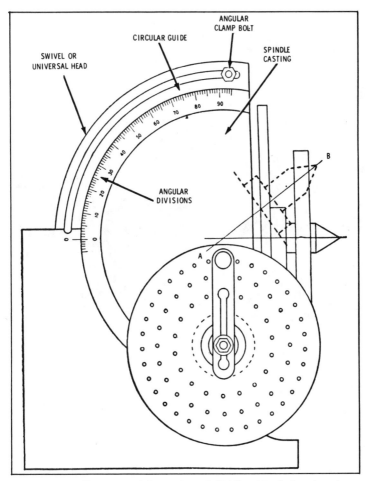

Figure 13-4 Diagram of the universal dividing head, showing the arrangement for swiveling the spindle (side view).

The universal dividing head is built for a greater range of work than the plain dividing head. For some kinds of tapered work, it is necessary to tilt the spindle at an angle to the table. The universal dividing head is designed for this kind of work. Angular graduations at the top of the housing serve as a guide to setting the spindle at any angle with reference to the horizontal. Figure 13-5 shows a milling operation in which the spindle is tilted at an angle.

Figure 13-5 Setup for the dividing head for milling cams.

(Courtesy Cincinnati Milacron Co.)

Helical Dividing Head

The *helical dividing head* differs from the plain and universal types in that the spindle of the head can be connected to the table lead screw by gears so that the work can be rotated as it is moved longitudinally by the table. The two movements are in a definite ratio that is determined by the combination of gears used. These gears are similar to the feed gears on a lathe.

The combination of the rotary and longitudinal movements causes the tool to cut a helix *AB*, as shown in Figure 13-6. The cutting tool begins at *A*, cutting the helix from *A* to *B* as the tool rotates in the direction indicated by *C*, while the table is moving the work in the direction indicated by *D*. The helix cut is either a right-hand or left-hand helix, depending on the direction in which the work is rotated.

The pitch of the helix depends on the rate of rotation of the work with respect to the movement of the table. Milling the flutes of a cast-iron rotor is shown in Figure 13-7.

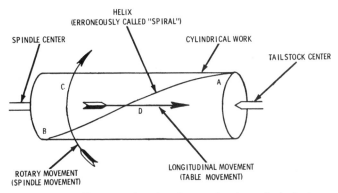

HELIX
(ERRONEOUSLY CALLED "SPIRAL")

SPINDLE CENTER

CYLINDRICAL WORK

TAILSTOCK CENTER

C

A

D

B

ROTARY MOVEMENT
(SPINDLE MOVEMENT)

LONGITUDINAL MOVEMENT
(TABLE MOVEMENT)

Figure 13-6   Diagram showing the production of a helix by a combination of rotary and longitudinal feeds.

Figure 13-7   Milling the flutes of a cast-iron rotor. *(Courtesy Cincinnati Milacron Co.)*

The basic parts of a helical dividing head are shown in Figure 13-8. A plain dividing head is driven by a gear train from the table feed screw, which converts it into a helical dividing head. A bevel gear (shown in dotted lines in Figure 13-3 on the shaft to which the worm is attached) is usually free to rotate on the worm shaft. The index plate is fastened to the hub of the bevel gear and is kept from rotating on the shaft by a stop pin in the housing. The index crank is attached to the worm shaft (see Figure 13-3). If the stop pin is withdrawn and the index pin is inserted in one of the holes in the index plate, the worm shaft and the bevel gear are locked together. Hence, any motion transmitted to the bevel gear is transmitted to the spindle through the gear train, index plate, index pin, index crank, worm, and worm wheel (see Figure 13-3).

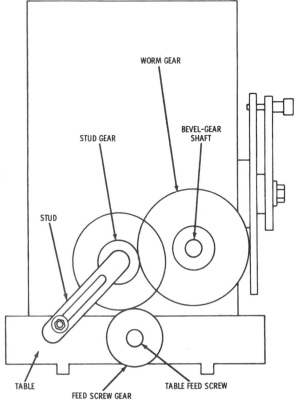

Figure 13-8 Diagram of a plain dividing head, showing the gear train used in cutting a helix.

The bevel gear (see Figure 13-3) meshes with another bevel-gear shaft (Figure 13-8). Spur gears of various sizes can be attached to the outer end of the bevel-gear shaft.

The bevel-gear shaft (which drives the spindle) is driven by the table feed screw through the gear train, which consists of the feed screw gear, two stud gears, and the worm gear (see Figure 13-8). The worm gear is actually a spur gear, but it is usually called the worm gear because it drives the worm in the driving head.

Various changes in the relative movements of the table feed screw and the spindle can be obtained by using gears of different sizes. Figure 13-8 shows only a single gear train. Sometimes an idler gear is interposed, as on an engine lathe when the selected gear diameters do not permit direct connection of the feed screw and the bevel-gear shaft.

Dividing Head and Driving Mechanism

The Cincinnati Milacron Co. manufactures dividing heads in 10-inch, 12-inch, and 14-inch sizes. These dividing heads are used extensively in milling spiral and helical gears, cams, etc. Motion is transmitted to the spindle in the same manner in all the different sizes of heads. Two types of driving mechanisms can be used on all heads:

> *Enclosed standard-lead driving mechanism*—Standard equipment on universal knee-and-column machines and available for plain and vertical machines
>
> *Short and long lead attachment*—Extra equipment

The dividing head is designed so that its spindle can be swiveled vertically. The spindle is housed in a swivel block. It can be swiveled to any angle from 5° below the horizontal to 5° beyond the vertical. Thus, bevel gears of any pitch angle can be milled, as well as many other kinds of work that require concentrically spaced slots or holes at an angle to the centerline of the workpiece.

Because of the splined shaft drive, the headstock of the dividing head is not required to be placed flush with the end of the table when set up with the driving mechanism. This kind of setup is required for milling scrolls and certain kinds of cams.

Setup of Dividing Head and Driving Mechanism

The following instructions for setting up the dividing head and driving mechanism should be followed in order:

1. The table of the milling machine and the bottoms of the dividing head and tailstock should be cleaned.

2. Clamp the headstock of the dividing head in the center slot of the table, in a location suitable for the work.

3. Check the spindle of the dividing head with a test bar and indicator to be certain that it is parallel to the table.

4. Depending on the length of the work, clamp the tailstock in proper position.

5. Align the tailstock center with the headstock center.

6. Center and align the cutter with the dividing head or head-stock center.

7. Lock the saddle in position.

8. Swing the table to the helix angle that is to be milled (universal machine only). If a universal milling attachment is used on a plain milling machine, swing it to the helix angle that is to be milled.

9. Lock the housing in position (universal machines only).

10. Withdraw the index plate stop. The index plate must be free to revolve with the index pin.

Note

The stop that engages the notches in the rim of the index plate should be engaged only when the dividing head is used without the driving mechanism.

11. Set up the change gears.

12. Set the sector of the index plate for the proper spacing.

13. Oil the dividing head and the change gears thoroughly.

Note

The use of the power rapid traverse for the table is not recommended when the driving mechanism is connected to the dividing head—especially when the head is equipped with a wide-range divider.

How to Select the Proper Change Gears

If you want to cut a helix having a 15¼-inch lead, consult the "Table of Leads" and find the lead that is nearest 15¼ inches—15.250 inches, in this instance (Figure 13-9). The change gears for this lead, 51, 18, 21, and 39, can be used because the lead obtained is near enough to the desired lead for practical purposes.

In some instances, the gear C is not listed for some leads. Then, only one intermediate gear is required, and a collar is used to replace gear C (Figure 13-10).

Lead of Spiral in Inches	A	B	C	D	Lead of Spiral in Inches	A	B	C
15.241	45	17	19	33	**15.966**	60	17	19
15.256	51	18	21	39	**15.983**	51	27	33
15.273	42	33	36	30	**16.000**	48	36	..

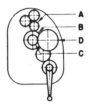

Figure 13-9   Typical "Table of Leads" for an enclosed driving mechanism.

Figure 13-10   Enclosed standard-lead driving mechanism. *(Courtesy Cincinnati Milacron Co.)*

Setting Up the Change Gears

Remove the bell-shaped cover on the apron at the right-hand end of the table. It is held in position by a large slotted-head screw. Then, place the change gears in the positions indicated for the desired lead, making certain that gears B and C are not interchanged. The gears X and Y determine whether the lead is in the right-hand or left-hand direction, as shown in Table 13-1.

Table 13-1 Determining Direction of Lead

Dividing Head	Right-Hand Helix	Left-Hand Helix
12 and 16 in. spiral	Remove gear Y. Reverse gear X.	Gears X and Y, as shown in Figure 13-10
	Gears X and Y, as shown in Figure 13-10	Remove gear Y. Reverse gear X.

Courtesy Cincinnati Milacron Co.

After completing the setup, move the table by means of the hand-feed crank before engaging the power feed. Remove the hand crank and keep the cover closed while the machine is in operation.

Summary

The dividing head is a device that can be used to rotate a piece of work through given angles, usually equal divisions of a circle. The types of dividing heads used on milling machines are plain, universal, and helical. Dividing heads also can be classified in different sizes.

In a plain dividing head, the spindle rotates about a horizontal axis. The universal dividing head differs from the plain head in that the spindle can be tilted or swiveled to any angular position in a vertical plane. The helical dividing head is different since the spindle of the head can be connected to the table lead screw by gears so that the work can be rotated as it is moved longitudinally by the table. The two movements are in a definite ratio that is determined by the combination of gears used.

Dividing heads are manufactured in 10-inch, 12-inch, and 14-inch sizes. These dividing heads are used extensively in milling spiral and helical gears, cams, and so on.

One thing to remember is that the use of the power rapid traverse for the table is not recommended when the driving mechanism is connected to the dividing head—especially when the head is equipped with a wide-range divider.

Review Questions

1. What is a dividing head?
2. What are the three dividing heads used?
3. What are the basic differences between the dividing heads?
4. The worm gear is keyed to the _____ so that the spindle turns with the wheel.
5. What is an index plate?
6. Where is the index pin located?
7. Describe the universal dividing head.
8. Describe the helical dividing head.
9. A helix can be produced by a combination of rotary and _____ feeds.
10. How do you set up the change gears?

Chapter 14

Indexing Operation

The dividing head (also called index head) is a device used to rotate a workpiece through given angles—usually equal divisions of a circle. *Indexing* is the operation in which the spindle of an index head is rotated through a desired angle by means of turning the index crank, which controls the interposed gearing.

An *indexing plate* is a plate perforated with variously spaced holes arranged in concentric circles and is used in a milling or similar machine for dividing work (such as spacing out teeth in gear cutting). An *index wheel* is a circular wheel or disc graduated around its circumference for indicating the angular measurements through which it has been moved. It is used on dividing engines for testing or for adjusting the feed on lathes.

In most dividing heads, 40 revolutions of the index crank are required for one revolution of a spindle (a ratio of 40:1). Some dividing heads are constructed with a 5:1 ratio.

Index Plates

A great variety of index plates are available for all indexing requirements. Usually, three index plates are provided with the dividing head, each plate consisting of six concentric circular rows of holes. Each circular row of holes is designated as an *index circle*. The circular row that contains 36 holes is referred to as the "36 circle." A typical set of index plates contains six circular rows of holes, as follows:

> Plate No. 1—15, 16, 17, 18, 19, 20
> Plate No. 2—21, 23, 27, 29, 31, 33
> Plate No. 3—37, 39, 41, 43, 47, 49

The dividing head manufactured by the Cincinnati Milacron Co. is equipped with one standard index plate that has a different series of holes on each side of the plate. The number of holes in each circular row of holes is as follows:

> First side—24, 25, 28, 30, 34, 37, 38, 39, 41, 42, 43
> Second side—46, 47, 49, 51, 53, 54, 57, 58, 59, 62, 66

One of three special index plates can be used to replace the standard plate. Each special index plate has 11 circular rows of holes on each side. Each circular row contains the number of holes, as shown in Table 14-1.

Table 14-1 Special Index Plates

Side	Number of Holes in Each Circular Row										
A	30	48	69	91	99	117	129	147	171	177	189
B	36	67	81	97	111	127	141	157	169	183	199
C	34	46	79	93	109	123	139	153	167	181	197
D	32	44	77	89	107	121	137	151	163	179	193
E	26	42	73	87	103	119	133	149	161	175	191
F	28	38	71	83	101	113	131	143	159	173	187

Divisions from 2 to 400,000 can be indexed directly on the "Wide Range Divider" manufactured by the Cincinnati Milacron Company. A large index plate, sector, and crank together with a small index plate, sector, and crank are used on the device. The crank on the small index plate operates through reduction gearing of 100:1 ratio. The ratio between the worm shaft and the spindle is 40:1.

Methods of Indexing
The dividing head can be used for several different methods of indexing. Index tables are provided by the manufacturer of the machine to aid in obtaining angular spacings and divisions.

Direct or Rapid Indexing
Direct indexing, also called *rapid indexing*, is used only on work that requires a small number of divisions (such as square and hexagonal nuts). In direct indexing, the spindle is turned through a given angle *without* the interposition of gearing. This is the simplest method of rotating the spindle through a given angle and is accomplished by turning the spindle by hand.

The index plate is fastened directly to the spindle, so that one complete revolution of the index plate rotates the spindle one complete revolution. Index plates of this type have only a few holes, but they should have as many holes as the number of divisions requires.

For example, an index plate with 24 holes can be used for any number of equal divisions divisible into 24. The divisions that are possible with a 24-hole index plate are as follows:

$$\frac{24}{24} = 1; \quad \frac{24}{12} = 2; \quad \frac{24}{8} = 3; \quad \frac{24}{6} = 4; \quad \frac{24}{4} = 6; \quad \frac{24}{2} = 12$$

An index plate can be used for any number of divisions that divides the number of holes in the plate equally. A plate with a larger number of holes than is required can be used if the number of holes is an exact multiple of the number of divisions required. In production work, it is safer to use a plate that has exactly the same number of holes as the required number because the possibility of making a mistake in indexing is reduced. Figure 14-1 shows the basic parts of an index plate that can be used for direct indexing.

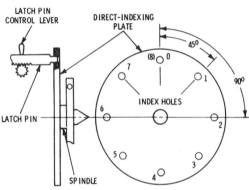

Figure 14-1 Basic diagram of an index plate, simplified to illustrate the principles of direct indexing.

As shown in Figure 14-1, the index plate (for the sake of simplicity) consists of eight index holes. Following are the possible divisions:

$$\frac{8}{8} = 1; \quad \frac{8}{4} = 2; \quad \frac{8}{2} = 4; \quad \frac{8}{1} = 8$$

Rotating the index plate eight, four, two, and one holes corresponds to angular movements of the plate and spindle of 360°, 180°, 90°, and 45°, respectively.

In all methods of indexing, it should be remembered that the hole from which the latch pin is disengaged when beginning to rotate the index plate is *not counted*; it is the "zero" hole. Actually, the number of spaces between the holes is counted, but it is the usual practice to count the number of holes—except for the beginning hole. The result is the same. The holes are counted, rather than the spaces, because it is much easier for the eyes to follow the holes when counting.

When milling a four-sided workpiece, such as a square nut, the first side is milled with the index pin in the "zero" hole; the second, fourth, and sixth index holes are used, since the remaining sides of the square nut are machined. As noted in Figure 14-1, the equal spacings—0–2, 2–4, and 4–6—each correspond to a 90° angular movement of the spindle. Similarly, in milling an eight-sided workpiece, the index plate is moved one hole for milling each side.

As mentioned, for simplicity, an index plate with only one circular row of eight holes was used in Figure 14-1, but index plates are provided that have a greater number of holes. For example, a standard index plate will have three concentric circular rows of holes consisting of 24, 30, and 36 holes, thereby increasing the range of the plate.

Plain Indexing

Plain indexing (sometimes called *simple indexing*) is an indexing operation in which the spindle is turned through a given angle *with* the interposition of gearing between the index crank and the spindle. This gearing usually consists of a worm on the index crankshaft that meshes with a worm wheel on the spindle. Several turns of the index crank are required for each rotation of the spindle—usually a ratio of 40:1.

The gearing arrangement provides a wide range of divisions or angular movements that are impossible on a direct-indexing plate. Because the index crank makes 40 revolutions to each revolution of the spindle, one revolution of the crank provides $\frac{1}{40}$ revolution of the spindle. Likewise, 5, 10, and 20 revolutions turn the spindle one-eighth, one-fourth, and one-half revolutions, respectively.

All dividing heads or index heads are provided with index plates that have circular rows containing a different number of holes, so almost any fractional turn of the index crank can be accomplished by using the correct index plate.

Use of the Dividing-Head Sector

On the Cincinnati Milacron Company dividing head, a sector (which is concentric with the index plate and crank) is used for plain indexing a fraction of a turn—especially when the procedure is repeated a number of times. The sector enables the operator to locate the index pin in the correct hole without counting the number of spaces that the pin must be moved forward for each division on the work (that is, after the sector has been set for the required number of spaces, as shown in Figure 14-2). The two arms of the sector are set apart at a distance that includes one hole more than the number of spaces that are to be indexed. This factor is sometimes overlooked and can be the source of error in setting up work on the dividing head.

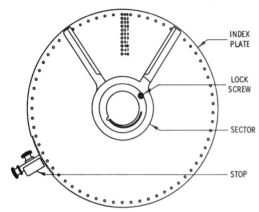

Figure 14-2 Diagram of dividing-head sector.

In setting the sector, the narrow edge of the left-hand sector arm should be placed in contact with the index pin. The index crank should be moved the required number of spaces and then dropped into the corresponding hole. Then, the right-hand sector arm should be placed against the pin and the sector arms tightened by means of a lock screw (see Figure 14-2).

The following example can be used to describe plain indexing. If the correct setting for a 23-tooth gear is desired, first consult the "Index Tables," and note that the 46-hole circle must be used for 23 divisions (Figure 14-3). Set the index plate so that the side with the 46-hole circle faces the index pin. Set the index pin in any hole in the 46-hole circle, and space the sector for 34 spaces (see Figure 14-3). Then, for each of the 23 divisions, rotate the index pin through one revolution of the crank, plus the sector spacing, or 34 spaces.

The index plate stop (see Figure 14-2) engages the notches in the index plate, keeping it from rotating. For a spiral or helical milling job, the index plate, sector, and crank rotate as a unit if the dividing head is connected to the driving mechanism. The index plate stop

No. of Divisions	Circle	Turns	Spaces	No. of Divisions	Circle	Turns	Spaces	No. of Divisions	Circle	Spaces	No. of Divisions	Circle	Spaces	No. of Divisions	Circle
2	Any	20	....	37	37	1	3	80	24	12	148	37	10	248	62
3	24	13	8	39	39	1	2	82	41	20	150	30	8	250	25
21	42	1	38	56	28	....	20	112	28	10	196	49	10	344	43
22	66	1	54	57	57	....	40	114	57	20	200	30	6	360	54
23	46	1	34	58	58	....	40	115	46	16	204	51	10	368	46
24	24	1	16	59	59	....	40	116	58	20	205	41	8	370	37
25	25	1	15	60	24	....	16	118	59	20	210	42	8	376	47
26	39	1	21	62	62	....	40	120	24	8	212	53	10	380	38
27	54	1	26	64	24	....	15	124	62	20	215	43	8	390	39
28	42	1	18	65	39	....	24	125	25	8	216	54	10	392	49

Figure 14-3 The correct circular row of holes to use for the desired number of divisions can be obtained from the index table.

must be disengaged for such a setup. The stop should be engaged only when the dividing head is not connected to the power drive, as when milling spur gears, bolt heads, etc. The stop serves as a safety precaution to prevent errors that would occur if the index plate were moved slightly while indexing.

The stop can also be used to reset work that has been removed for inspection purposes. First, reset the work in an approximate relationship with the cutter. Withdraw the index plate stop, and with the index pin engaged, rotate the crank a sufficient amount to position the work accurately. Reengage the stop in the notches on the rim of the index plate. Two inches of the circumference of the index plate are notched; the notches have a pitch of 0.060 inch. Thus, moving the index plate one notch is equivalent to rotating the work 1/18, 460 revolution.

How to Calculate Indexing on the Standard Driving Head
The following rules and example illustrate the procedure for obtaining the maximum number of settings for indexing. Remember, the ratio between the worm and worm wheel on the Cincinnati dividing head is 40:1.

1. Divide 40 by the number of divisions required, to give the number of turns or fractional turn of the index crank.

2. If a fraction of a turn is required, the denominator (lower part of the fraction) represents the circle to be used; the numerator

(upper part of the fraction) represents the number of spaces in the circle that must be passed over by the index pin.

3. Reduce the fraction to its lowest terms. Multiply both the numerator and denominator by the same number until the denominator equals the number of holes in one of the circles on the index plate.

For example, say you want to calculate all the indexing circles that can be used for three divisions. The following formula can be used:

$$t = \frac{N}{D}$$

in which t = number of completed turns and/or fraction of a turn of the index crank.

N = number of turns of the index crank for each revolution of the dividing-head spindle or workpiece. This is equal to 40 turns of the dividing head.

D = number of divisions required in the workpiece.

Thus, $N = 40$ and $D = 3$. Therefore, from this formula:

$$t = \frac{40}{3}$$

$t = 13\frac{1}{3}$ revolutions

One-third of a revolution can be obtained by rotating the index pin over one space in a three-division circle (Rule 2). As a three-hole index circle is not available, an index circle having a number of holes that can be divided into three equal divisions must be used. For example, 8 spaces in the 24-hole circle ($\frac{8}{24} = \frac{1}{3}$), 10 spaces in the 30-hole circle ($\frac{10}{30} = \frac{1}{3}$), and so on, can be used. The one-third turn can be obtained in each of the following index circles:

- $\frac{1}{3} \times \frac{8}{8}$ = $\frac{8}{24}$ or 8 spaces in the 24-hole circle.
- $\frac{1}{3} \times \frac{10}{10}$ = $\frac{10}{30}$ or 10 spaces in the 30-hole circle.
- $\frac{1}{3} \times \frac{13}{13}$ = $\frac{13}{39}$ or 13 spaces in the 39-hole circle.
- $\frac{1}{3} \times \frac{14}{14}$ = $\frac{14}{42}$ or 14 spaces in the 42-hole circle.
- $\frac{1}{3} \times \frac{17}{17}$ = $\frac{17}{51}$ or 17 spaces in the 51-hole circle.
- $\frac{1}{3} \times \frac{18}{18}$ = $\frac{18}{54}$ or 18 spaces in the 54-hole circle.

- $\frac{1}{3} \times {}^{19}/_{19} = {}^{19}/_{57}$ or 19 spaces in the 57-hole circle.
- $\frac{1}{3} \times {}^{22}/_{22} = {}^{22}/_{66}$ or 22 spaces in the 66-hole circle.

The formula can also be used for indexing *more than 40 divisions*. For example, say you want to index 152 divisions:

$$t = \frac{40}{152}$$

Since the index plate does not have a circle containing 152 holes, it is necessary to transform the fraction into an equivalent fraction with the denominator equal to the number of holes in one of the circles of the index plate. Since the index plate does contain a 38-hole circle, the fraction (${}^{40}/_{152}$) can be reduced to ${}^{10}/_{38}$ by *dividing* both the numerator and the denominator by 4 as follows:

$$\frac{40 \div 4}{152 \div 4} = \frac{10}{38}$$

Thus, the 38-hole circle can be used, and the index pin must be moved over a series of 10 holes for each of the 152 divisions into which the work is to be divided.

If you want to index *less than 40 divisions* (33 divisions, for example), the fraction then becomes ${}^{40}/_{33}$ ($t = {}^{40}/_{33}$). The index plate does not contain a 33-hole circle; also, both an 11-hole circle and a 3-hole circle are lacking. Since these are the only numbers that can be divided into 33, the transformation to a fraction that can be used must be accomplished by *multiplying* both the numerator and denominator by the same number, rather than by dividing, as in the above example. As the index plate contains a 66-hole circle, both the numerator and denominator can be multiplied by 2 as follows:

$$\frac{40 \times 2}{33 \times 2} = \frac{80}{66}$$

In the equivalent fraction ${}^{80}/_{66}$, the denominator (66) represents the index circle and the numerator (80) represents the number of holes over which the index pin must pass for each division. As the 66-hole circle does not contain 80 holes, the pointer must make one complete turn plus 14 more holes ($80 \div 66 = 1{}^{14}/_{66}$).

Calculation of indexing can be illustrated in a practical milling problem. Calculate the indexing for milling a hexagonal nut.

The gearing is in 40:1 ratio, so apply the following formula ($t = N/D$):

$$t = \frac{40}{6} = 6\tfrac{4}{6} \text{ revolutions}$$

The crank can be rotated exactly $\tfrac{4}{6}$ of a revolution by selecting an index plate with the correct number of holes in one of its index circles. By reducing the fraction ($\tfrac{4}{6}$) to its lowest terms ($\tfrac{2}{3}$), an index circle having a number of holes equally divisible by the denominator (3) can be selected.

Assuming that the index plate has index circles with 37, 39, 41, and 49 holes, the 39-circle can be selected (by inspection) because it is the only circle that has a number of holes that is equally divisible by 3 (that is, $39 \div 3 = 13$). Thus, moving the index crank 13 holes is equivalent to the following:

$$\frac{13}{39} = \frac{1}{3} \text{ revolution of the index crank}$$

Because the index crank must be moved $\tfrac{2}{3}$ of a revolution, the crank must be moved (13×2) or 26 holes for the fractional part of the $6\tfrac{2}{3}$ revolutions. Therefore, the index crank is turned six complete revolutions plus 26 holes for each indexing, when milling the hexagonal nut. Before the milling operation is begun, the crank must be adjusted radially so that the index pin will register with the holes in the 39-hole index circle.

As mentioned, to avoid counting the number of holes for the fractional turn, the sector can be adjusted for the 26 holes. For review, set one arm of the sector to touch the index pin, and set the other arm so that it barely uncovers the 26th hole from the pin—*not* counting the hole engaged by the pin. Then, clamp the sector arms in position.

Compound Indexing

The *compound indexing operation* is a method of obtaining a desired spindle movement by first turning the index crank to a definite setting as in plain indexing and then turning the index plate itself to locate the crank in the correct position. Normally, the index plate is held stationary by a stop pin, which engages one of the index holes. When the stop pin is disengaged, the index plate can be rotated. Compound indexing can be used to obtain divisions that are beyond the range of the plain indexing system.

Occasionally, none of the index plates provided with the machine will have a number of holes that is suitable for providing the correct number of holes for the required fractional turn of the index crank; this can be overcome by compound indexing. Two kinds of compound indexing can be used, according to the relative movements of the index crank and index plate:

- Positive compounding
- Negative compounding

Positive Compounding

To illustrate compound indexing, assume that the index plate has both a 19-hole and a 20-hole circle. Move the index crank one hole in the 19-hole circle. Disengage the stop pin, and rotate the index plate one hole in the *same* direction in the 20-hole circle; this is positive compounding (Figure 14-4).

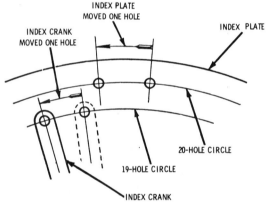

Figure 14-4 Basic diagram of an index plate, showing positive compounding.

The two movements cause the worm to rotate:

$$\frac{1}{19} + \frac{1}{20} = \frac{20 + 19}{380} = \frac{39}{380} \text{ of a revolution}$$

As the worm–spindle ratio is 40 to 1,

$$\frac{39}{380} \div 40 = \frac{39}{15,200} \text{ of a revolution}$$

Negative Compounding
Note that in negative compounding, the movements are in opposite directions (Figure 14-5). Move the index crank one hole in the 19-hole circle. Disengage the stop pin and rotate the index plate one hole in the *opposite* direction in the 20-hole circle. The resultant rotation of the worm is as follows:

$$\frac{1}{19} - \frac{1}{20} = \frac{1}{380}$$

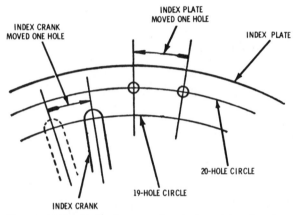

Figure 14-5 Basic diagram of an index plate, showing negative compounding.

The worm–spindle ratio is 40 to 1; the spindle movement, therefore, is

$$\frac{1}{380} \div 40 = \frac{1}{15,200}$$

To index work that requires divisions such as a 69-hole circle, for example, the crank should be turned:

$$\frac{1}{69} \times 40 = \frac{40}{69} \text{ of a revolution}$$

If the index plate contained a 69-hole circle, it would only be necessary to move the crank 40 holes to obtain one spindle revolution. However, in the absence of a 69-hole plate, the same result can be obtained by compounding, using an index plate with both 23-hole and 33-hole circles.

In compound indexing with these index circles, first move the crank counterclockwise 21 holes in the 23-hole circle. Then, withdraw the stop pin and move the index plate clockwise 11 holes in the 33-hole circle. The resulting movement of the crank from its original position is

$$\frac{21}{23} - \frac{11}{33} = \frac{693 - 253}{759} = \frac{440}{759} = \frac{40}{69}$$

As the ratio between the crank and spindle is 40 to 1,

$$\frac{40}{69} \times \frac{1}{40} = \frac{1}{69} \text{ revolution of the spindle}$$

Determining Which Index Circles to Use

The procedure for determining the index circles that can be used can be illustrated for the previous example (1/69 revolution of the spindle; 40 to 1 ratio), as follows:

1. Resolve into factors the number of divisions required.

$69 = (23 \times 3)$

2. For trial and error, choose an index plate. Arbitrarily select two index circles (for example, a 23- and a 33-hole circle).

3. Subtract the number of holes in the smaller index circle from the number of holes in the larger index circle.

$33 - 23 = 10$

4. Factor the difference.

$10 = (2 \times 5)$

5. Place the two factors (from 1 and 4) *above* a horizontal line.

$(23 \times 3)(2 \times 5)$

6. Factor the number of turns of the crank required for one revolution of the spindle.

$40 = (2 \times 2 \times 2 \times 5)$

7. Factor the number of holes for each of the trial index circles.

$23 = (23 \times 1)$
$33 = (3 \times 11)$

8. Place these three sets of factors (from 6 and 7) *below* the horizontal line.

$$\frac{(23 \times 3) \ (2 \times 5)}{(2 \times 2 \times 2 \times 5) \ (23 \times 1) \ (3 \times 11)}$$

9. Cancel the equal factors both above and below the horizontal line.

$$\frac{23 \times 3 \times 2 \times 5}{2 \times 2 \times 2 \times 5 \times 23 \times 1 \times 3 \times 11} = \frac{1}{2 \times 2 \times 1 \times 11}$$

If all factors above the line cancel, the two selected trial index circles can be used. If all factors above the line *do not* cancel, two other index circles must be selected for trial and error, and the procedure must be repeated.

10. To obtain crank movement in a forward direction and plate movement in a reverse direction, multiply the uncanceled factors below the horizontal line.

$2 \times 2 \times 1 \times 11 = 44$

11. Thus, the indexing number is 44. This means that to move the spindle $\frac{1}{69}$ revolution, the index crank must be turned forward 44 holes in the 23-hole circle, and the index plate must be turned 44 holes in the reverse direction in the 33-hole circle (step 3). The same result can be obtained by making the forward movement in the 33-hole circle and the reverse movement in the 23-hole circle.

This example illustrates negative compounding. Plus (+) and minus (−) symbols are used to indicate the forward and reverse directions of movement in the indexing tables.

In compound indexing, the choice of correct index circles (step 2) usually cannot be solved on the first attempt, as in the example. Normally, two or more selections are necessary before all the factors above the horizontal line can be canceled (step 9).

Compound indexing should not be used when the required divisions can be obtained by simple indexing, because there is a greater possibility for making an error. For this reason, compound indexing has been replaced to a large extent by the differential method. However, knowledge of compound indexing can be valuable for the machinist.

Sometimes the plain and compound indexing systems can be combined to advantage in an operation such as gear cutting. Every other tooth can be cut by plain indexing, and then the spindle can be positioned to locate the cutter in the center of spaces already cut. Then, the remaining spaces are cut by plain indexing as before.

Differential Indexing

In this method of indexing, the spindle is turned through a desired division by manipulating the index crank. The index plate is rotated, in turn, by proper gearing that connects it to the spindle. As the crank is rotated, the index plate also rotates a definite amount, depending on the gears that are used. The result is a differential action of the index plate, which can be either in the same direction or in the opposite direction in relation to the direction of crank movement, depending on the gear setup. As motion is a relative matter, the actual motion of the crank at each indexing is either greater or less than its motion relative to the index plate.

In compound indexing, the index plate is rotated manually, with a possibility of error in counting the holes. This is avoided in differential indexing; therefore, chances for error are greatly reduced.

Usually, determination of the gears between the index plate and the spindle can be accomplished by means of an index table that accompanies the machine. The table provides data for both plain and differential gearing. Thus, it can be determined quickly whether the latter method can be used.

In the differential indexing operation, the index crank is moved relative to the index plate in the same circular row of holes in a manner that is similar to plain indexing. Since the spindle and index plate are connected by interposed gearing, the index plate stop pin on the rear of the plate must be disengaged before the plate can be rotated.

In the gearing hookup, the number of idlers determines whether the plate movement is *positive* (in the direction of crank movement) or *negative* (opposite the direction of crank movement). The gear arrangements are as follows:

- *Simple*—The use of one idler provides positive motion, or the use of two idlers provides negative motion to the index plates.
- *Compound*—The use of one idler provides negative movements, or the use of two idlers provides positive movement.

In general, the spindle rotates by means of the worm and worm wheel gearing, as the crank is turned. The index plate is rotated by the gearing between the spindle and the plate. The direction of rotation is either positive or negative, depending on the gear hookup. The total motion or movement of the crank in indexing is equal to its total movement in relation to the index plate—that is, the sum of its positive motion and its negative motion.

In *simple differential indexing* (Figure 14-6), the gear on the worm shaft is the driver, and the gear on the spindle is the driven gear. The ratio between the number of teeth in each gear determines the spacing between the divisions. The number of teeth in the idler is not important, because the idler is used only to connect the other two gears and to cause the gear on the spindle to rotate in the desired direction. Two idlers are often used.

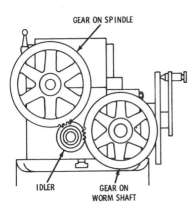

GEAR ON SPINDLE

IDLER

GEAR ON
WORM SHAFT

Figure 14-6 Gearing diagram for simple differential indexing.

In *compound differential indexing* (Figure 14-7), the idler is also used to cause the gear on the spindle to rotate in the desired direction. Again, the number of teeth in the idler is not important, but the other gears must have the correct number of teeth for the desired spacing. The gear on the worm shaft and the second gear on the stud are driver gears, and the first gear on the stud and the gear on the spindle are driven gears.

To select the correct change gears, it is necessary to find the required gear ratio between the spindle and the index plate. The correct gears can then be determined. The following formulas can be used to determine these gears, in which

N is the number of divisions required,

H is the number of holes in the index plate,

n is the number of holes taken at each indexing,

V is the ratio of gearing between the index crank and the spindle,

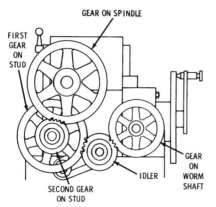

Figure 14-7 Gearing diagram for compound differential indexing.

X is the ratio of the train of gearing between the spindle and the index plate,

S is the gear on the spindle (driven),

G_1 is the first gear on the stud (driven),

G_2 is the second gear on the stud (driver), and

W is the gear on the worm shaft (driver).

If HV is larger than Nn, then

$$X = \frac{HV - Nn}{H}$$

If HV is less than Nn, then

$$X = \frac{Nn - HV}{H}$$

For simple gearing,

$$x = \frac{S}{W}$$

For compound gearing,

$$x = \frac{SG_1}{G_2W}$$

Angular Indexing

The operation of rotating the spindle through a definite angle (in degrees) by turning the crank is called *angular indexing*. Sometimes, instead of specifying the number of divisions or sides required for the work to be milled, a given angle, such as 20° or 45°, may be specified for indexing.

The number of turns of the index crank required to rotate the spindle 1° must first be established to provide a basis for rotating the spindle through a given angle. Usually, 40 turns of the index crank are required to rotate the spindle one complete revolution (360°). Thus, one turn of the crank equals $360 \div 40 = 9$ degrees, or $\frac{1}{9}$ turn of the crank rotates the spindle 1°. Accordingly, to index one degree, the crank must be moved as follows:

- On a 9-hole index plate, 1 hole.
- On an 18-hole index plate, 2 holes.
- On a 27-hole index plate, 3 holes.

As an example, on an 18-hole index plate, calculate the crank movement needed to index 35°.

Since one turn of the crank equals 9°, $35 \div 9 = 3\frac{8}{9}$ turns of the crank for 35°. On an 18-hole plate, $\frac{1}{9}$ turn equals $18 \div 9$, or 2 holes, and $\frac{8}{9}$ turn equals 2×8, or 16 holes. Therefore, to index 35° on an 18-hole plate, three turns of the crank plus 16 holes on the plate are required.

It should be noted that

- One hole in the 18-hole circle equals $\frac{1}{2}°$.
- One hole in the 27-hole circle equals $\frac{1}{3}°$.
- Two holes in the 18-hole circle equals 1°.
- Two holes in the 27-hole circle equals $\frac{2}{3}°$.

Say you want to calculate the indexing for 15 minutes (15'). The calculation procedure is as follows:

One turn of the crank $= 9°$

$9 \times 60' = 540'$

$15' = \frac{15}{540}$ of one turn

Compound Angular Indexing

When the index crank is moved one hole in the 27-hole circle, the spindle rotates 20 minutes (20'); when the crank is moved one hole in the 18-hole circle, the spindle rotates 30 minutes (30'). The compound method can be used to index angles accurately to 1 minute (1'), as follows:

1. Place a 27-hole plate outside a 20-hole plate so that any two holes register, and fasten them together in position.

2. Turn the plates clockwise three holes in the 20-hole circle. Then, turn the crank counterclockwise four holes in the 27-hole circle. Thus, the total of these movements is a resultant spindle movement of exactly 1 minute (1') in a clockwise direction.

Block Indexing

This is sometimes called *multiple indexing* and is adapted to gear cutting. In this operation, the gear teeth are cut in groups separated by spaces. The work is rotated several revolutions by the spindle while the gear teeth are being cut.

For example, when cutting a gear that has 25 teeth, the indexing mechanism is geared to index four teeth at the same time. During the first revolution, six widely separated spaces are cut. During the second revolution, the cutter is placed one tooth behind the previously milled spaces. On the third indexing, the cutter drops behind still another tooth. In this example, the work is revolved four times to complete the gear.

The chief advantage of block indexing is that the heat generated by the cutter (especially when cutting cast iron gears with coarse pitch) is distributed more evenly around the rim of the gear. Thus, distortion caused by local heating is avoided, and higher speeds and feeds can be used.

Summary

An indexing plate is a plate perforated with variously spaced holes arranged in concentric circles and is used in a milling or similar machine for dividing work (such as spacing out teeth in gear cutting). An index wheel is a circular wheel or disc graduated around its circumference for indicating the angular measurements through which it has been moved. It is used on dividing engines for testing or for adjusting the feed on lathes.

A great variety of index plates are available for all indexing requirements. The dividing head can be used for several different methods of indexing. Index tables are provided by the manufacturer of the machine to aid in obtaining angular spacings and divisions. Direct indexing, also called rapid indexing, is used only on work that requires a small number of divisions (such as square and hexagonal nuts). In direct indexing, the spindle is turned through a given angle without the interposition of gearing.

Plain indexing is an indexing operation in which the spindle is turned through a given angle with the interposition of gearing between the index crank and the spindle. The gearing arrangement provides a wide range of divisions or angular movements that are impossible on a direct-indexing plate.

Compound indexing is a method of obtaining a desired spindle movement by first turning the index crank to a definite setting as in plain indexing and then turning the index plate itself to locate the crank in the correct position. There is both positive and negative compounding.

Differential indexing is a method of indexing whereby the spindle is turned through a desired division by manipulating the index crank. The index plate is rotated, in turn, by proper gearing that connects it to the spindle.

Compound angular indexing and block indexing are also available. Block indexing is sometimes called multiple indexing. The chief advantage of block indexing is that the heat generated by the cutter (especially when cutting cast iron gears with coarse pitch) is distributed more evenly around the rim of the gear. Thus, distortion caused by local heating is avoided, and higher speeds and feeds can be used.

Review Questions

   1. What is compound indexing?
   2. What is angular indexing?
   3. What is block indexing?
   4. What is an indexing plate?
   5. What is an indexing wheel?
   6. What is the purpose of the indexing operation?
   7. What is rapid indexing?
   8. What is plain indexing?
   9. What formula do you use to calculate indexing on the standard driving head?
  10. What is differential indexing?

Chapter 15

Shapers

The *shaper* is a metal-removing machine. The cutting tool is moved in a horizontal plane by a ram having a reciprocating motion, and it cuts only on the forward stroke. The work is usually held in a vise bolted to a boxlike table that can be moved either vertically or horizontally.

The size of a shaper is determined by the size of the largest cube that can be machined on it. The shaper was intended originally to produce only plane surfaces—angular, horizontal, or vertical—but it can be used to produce either concave or convex surfaces (Figure 15-1).

Figure 15-1   A 7-inch bench shaper. *(Courtesy South Bend Lathe, Inc.)*

There are two types of shapers—the "crank" and the "geared" shapers. The crank type is the most commonly used and is made in either standard or universal types. The shaper is used for machining small parts.

Basic Construction

The basic diagram of a typical shaper is shown in Figure 15-2. The diagram illustrates the general assembly of parts and their order.

Base

The *base* of the machine supports the column or pillar that supports all the working parts. An accurately machined dovetail forms a bearing slide for the ram, and the front of the column is machined to provide slides (or ways) for the up-and-down movement of the crossrail.

Crossrail

The *crossrail* is a heavy casting attached to the base by means of gib plates and bolts. The top surface of the crossrail is accurately machined to provide a smooth surface for the saddle as it is fed back and forth. The crossrail contains the table elevating and traversing mechanisms.

Saddle

The *saddle* is gibbed to the crossrail and supports the table. If the table is removed, the work can be bolted or clamped to the T-slots in the front of the saddle.

Table

The *table* is boxlike in construction and is bolted to the saddle. Either the work or the swivel for holding the work can be clamped or bolted to T-slots in the top and sides of the table. Some tables are constructed to swivel in a vertical plane on the saddle, or they can be tilted up and down. When a shaper is equipped with this kind of table, it is known as a *universal shaper*. The table can be moved horizontally by means of either a rapid traverse hand-feed or the automatic power feed, and it can be adjusted vertically for different thicknesses of work by means of the elevating screw.

The table feed mechanism consists of the pawl and ratchet mechanism, a rocker screw for regulating the amount of feed for each return stroke of the ram, and the cross-feed arm, which connects the rocker arm to the ratchet drive and lead screw. As the ratchet screw moves back and forth, the cross-feed arm moves with it, actuating the pawl and ratchet on the end of the feed screw. The crossrail holds the feed screw, which passes through a bronze nut in the back of the saddle. Any movement of the ratchet by the rocker screw causes the feed screw to move the table a definite distance.

A crank fits the squared end of the cross-feed screw. The table can be fed just as fast as the operator desires to turn the crank, or a

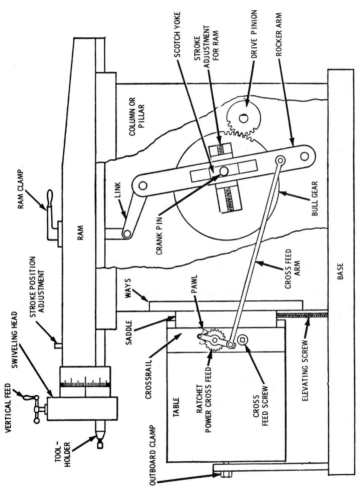

Figure 15-2 Basic diagram of a typical shaper.

lever that controls a high-speed power traverse mechanism can be used to move the table in either direction in relation to the operator. An eccentric slot in the bull gear provides movement for the automatic table feed. The eccentric slot makes one revolution for each revolution of the bull gear and transmits motion by means of a shoe and a series of lever arms to the rocker screw and cross-feed arm, so that the cross-feed arm moves backward and forward in each revolution of the bull gear.

The automatic power feed is engaged by first feeding the work to the cutting tool by hand-feed. The pawl is turned to engage the teeth on the ratchet feed screw. Since the power feed mechanism is always in motion while the shaper is running, the pawl will be moved back and forth, driving the ratchet in one direction or the other, depending on how the pawl contacts the ratchet. On the return stroke of the ram, the feed screw is moved, causing the table to be fed toward the cutting tool in preparation for the cutting stroke.

Ram

The *ram* is a strong and rigid casting that is actuated back and forth horizontally in the dovetail slide by means of the rocker arm and the crankpin. The ram contains a stroke-positioning mechanism and the downfeed mechanism.

The toolhead slides in a dovetail at the front of the ram by means of T-bolts. It can swivel from 0° to 90° in a vertical plane. The toolhead can be raised or lowered by hand-feed for vertical cuts on the workpiece.

Bull Gear

The *bull gear* is mounted on the column and is driven by a pinion that is moved by the speed-control mechanism. A radial slide, carrying a sliding block into which the crankpin fits, is anchored to the center of the bull gear. The position of the sliding block is controlled by a small lead screw connected to the operator's side of the shaper by means of bevel gears and a squared-end shaft. The location of the sliding block with respect to the center of the bull gear governs the length of stroke of the ram. The farther apart the two centers are located, the greater the length of stroke.

Shaper Operations

Several factors can affect the efficient operation of the shaper. Some of these factors are selection of the cutting tool, holding the work, adjustment of the work, adjustment of the stroke, and selection of proper feed and cutting speed.

Cutting-Tool Selection

The shank of the cutting tool should be in a plane perpendicular to the line of motion of the cutting tool. Therefore, the clearance angle should remain constant. The clearance angles on cutting tools for the shaper are not as large as those on cutting tools for the lathe. Also, the shaper requires a shorter cutting tool than the lathe, so the tendency to dig into the work is avoided.

Cutting tools that are commonly used on the shaper are shown in Figure 15-3. All these tools are available as either straight or bent cutting tools. Roughing, finishing, side roughing, and side finishing tools are most commonly used. Special toolholders with inserted-blade cutting tools can be used effectively on the shaper. Various special cutting tools are also used to perform certain operations. For example, one of these special cutting tools is used for cutting T-slots.

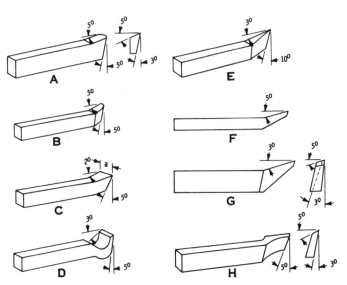

Figure 15-3 Cutting tools commonly used on the shaper are (A) right-hand roughing, (B) round nosed, (C) squared nosed, and (D) goosenecked. (E) and (F) are views of a right-hand down-cutting side tool; (H) is a left-hand side-facing tool.

Holding the Work

As shaper operations are concerned primarily with small work-pieces, the work is usually held in either a vise or a chuck. These holding devices are auxiliary to the table or machine.

Work Bolted Directly to the Table

It is important to place the stops (T-bolts) properly. If the work were laid on the table without the stops to resist the thrust of the cutting tool on the cutting stroke, the cutting tool would "push" the work rather than cut it.

Large pieces of work that cannot be held in a vise should be fastened to the table. In fastening the work to the table, an important consideration is to tighten the clamp bolts no more than is necessary to hold the work firmly on the table to avoid distorting or springing the work. Since the cutting tool does not tend to elevate the workpiece, extreme tightening of the bolts is not necessary.

T-bolts having a head of correct size to fit the table slots are commonly used. A complete bolt assembly consists of the bolt, nut, and washer (Figure 15-4). The bolt should have a thread of sufficient length for work of various thicknesses.

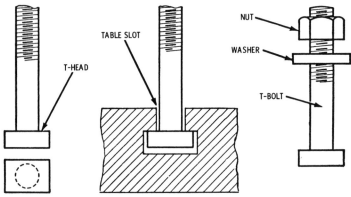

Figure 15-4 T-head bolt (left) and detail of table showing T-head bolt in the T-slot (center). T-head assembly consisting of bolt, washer, and nut is shown (right).

Fastening with Clamps

A complete clamping unit consisting of the bolt assembly, clamp, and fulcrum block is shown in Figure 15-5. In a clamping arrangement, the lever principle should always be considered. Thus, the work should be positioned on the table close to the table slot so that the bolt will be as far from the fulcrum block as possible. That is, the bolt should be nearer the work than the block (see Figure 15-5). Correct and incorrect methods of clamping are shown in Figure 15-6. It is important that the machinist remember to place

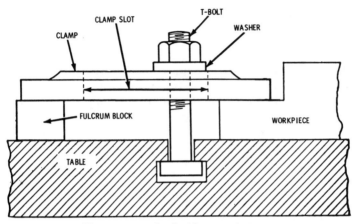

Figure 15-5   Clamping unit for holding workpiece.

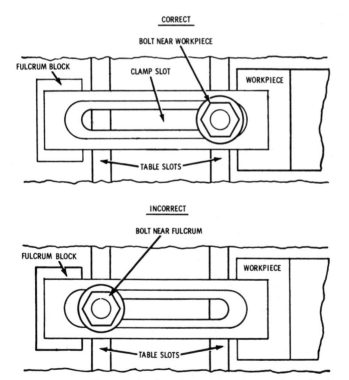

Figure 15-6   Correct (top) and incorrect (bottom) methods of setting up a clamping unit.

the bolt near the work and to tighten the bolts just enough to anchor the work, but not enough to spring the work.

If a number of clamps are used, turn all the bolts until the work is fastened lightly. Then, tighten the bolts in a staggered sequence to distribute the stresses brought onto the work by clamping (Figure 15-7). It is also important to select a fulcrum block of the same thickness (or height) as that of the work being clamped to the table. This gives the clamp an even bearing on the work, and the work will be held more securely (Figure 15-8).

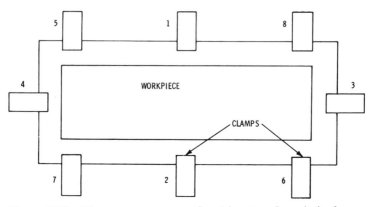

Figure 15-7 The proper sequence for tightening clamp bolts for proper distribution of clamping stresses.

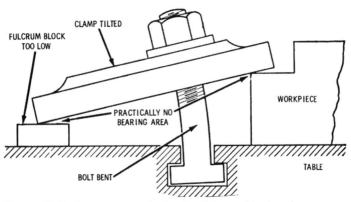

Figure 15-8 Importance of selecting fulcrum blocks of proper height to hold the work securely.

Stop Pins

Two types of stop pins are the *hole type* and the *inclined slot type* (Figure 15-9). A stop pin can be anchored in a hole or slot in the table. The screw can be forced directly against the work, or it can be used to force another device (such as a toe dog) against the work.

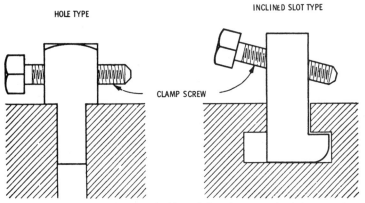

HOLE TYPE

INCLINED SLOT TYPE

CLAMP SCREW

Figure 15-9   The hole-type (left) and the slot-type (right) stop pins.

Stop Pins and Toe Dogs

A toe dog is a holding device that is similar in shape to a center punch or cold chisel, and it is designed to be forced against the work by a stop pin. Stop pins are used in combination with toe dogs to hold thin work. The pointed and blade types of toe dogs are shown in Figure 15-10, and their application is shown in Figure 15-11. Stop pins of either the hole type or the slot type can be inserted in the table on each side of the work, and the dogs can be forced against the work by tightening the stop-pin screws. Note that the end of the stop pin projects into the short bore in the end of the toe dog, connecting the two parts (see Figure 15-10 and Figure 15-11).

The work is pressed against the table because of the angular position of the dogs. However, the angle between the screw and the dog should not be too great. The work should be anchored by an ample number of combined stop pin and toe dog units so that they can take the thrust and secure the work to the table.

Stop Pins and Table Strip

A table strip can be used in combination with stop pins to hold work that is thick enough to project above the strip and stop-pin screw

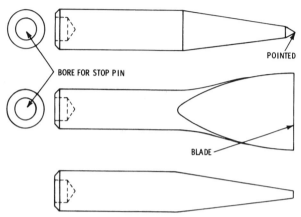

Figure 15-10   The pointed (top) and the blade (center and bottom) types of toe dogs.

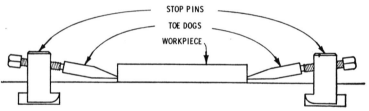

Figure 15-11   Application of stop pins and toe dogs for holding thin work.

(Figure 15-12). The table strip has a tongue on one side that fits in the table slot. This positions the work parallel to the travel of the cutting tool. A method of fastening the work by means of stop pins and table strip is shown in Figure 15-13. The table strip is held in the slot by two T-bolts. The tongue side of the strip projects downward into the slot. The stop pins are placed in the table holes, and the pins are turned firmly against the work. An adequate number of stops for taking the thrust of the cutting tool should be provided.

Braces for Tall Castings
Workpieces that project some distance above the table can be machined with the aid of braces, which are used to hold the work and to take the thrust of the cutting tool (Figure 15-14). The brace is used to aid the clamp (lower stop) in resisting the tendency of the casting to move upward in a circular path. Thus, the work is

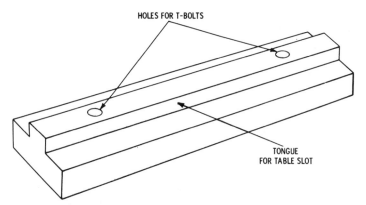

Figure 15-12 Bottom view of the table strip, showing the tongue and the bolt holes.

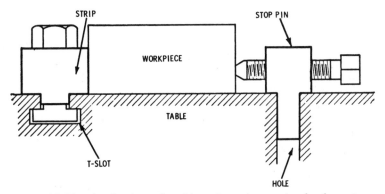

Figure 15-13 Application of a table strip and stop pins for fastening a workpiece.

stop-anchored on both the upper and the lower flanges, which provides a rigid fastening for the workpiece. The number of braces required depends on the size and shape of the workpiece. This also applies to the number of clamps and stop pins required.

Holding Circular Workpieces

If a length of shafting is to be splined or key seated, it must be fastened rigidly to the table, as shown in Figure 15-15. In this setup, a special table strip with an inclined edge A and a wedge block are used to hold the workpiece. As with an ordinary table strip, the

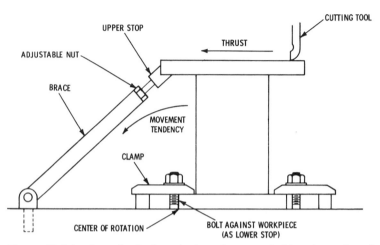

Figure 15-14 A method of employing a brace to hold a piece of work that projects above the table.

tongue provides parallelism, and a wedge block *B* with a distance piece and a stop pin completes the setup.

The axis *C* of the stop pin should be higher than the center *D* of the workpiece. If these conditions are provided, the workpiece will be clamped firmly to the table. If the axis of the stop pin were placed below the center of the workpiece, the wedge block would rotate.

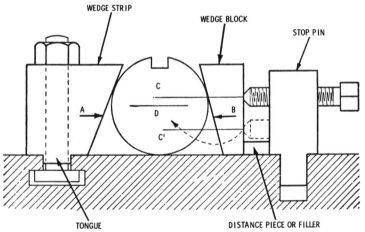

Figure 15-15 Setup for holding circular or round pieces of work.

On production work, the special holding devices, such as the strip and block, must be designed to provide the most efficient clamping of the workpiece. A stop should always be included to resist the thrust of the cutting tool.

Another method of clamping a shaft uses the table slot instead of a strip to provide parallelism (Figure 15-16). This method can be used on tables having transverse slots. Here again, the number and kind of holding devices for a given setup are determined by the size, shape, and any other special characteristics of the workpiece.

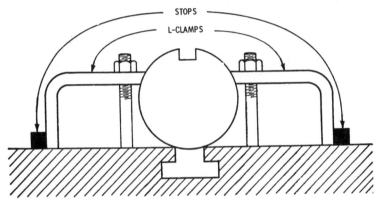

Figure 15-16 Setup in which L-clamps and stops are used to clamp a shaft. The table slot provides parallelism.

Irregularly Shaped Work Bolted Indirectly to the Table

Numerous fixtures can be used to set up castings having odd shapes for machining. Any device that can be attached or "fixed" to the table for holding and positioning the workpiece is considered to be a fixture. The angle plate is a fixture. One side is bolted (fixed) to the table, and the workpiece is bolted to the other side (Figure 15-17).

Fixtures are valuable aids in production work, as the successive castings not only can be held in position for machining but also can be held in the same position, when bolted to the fixture, as the previous ones. If the casting projects from the angle plate, provision can be made for supporting it on wedge blocks to prevent springing. The wedge blocks should not be tightened enough to spring the casting, but they should be tightened enough to provide support. Springing can be prevented by checking the setup with a surface gauge. Note that the angle plate acts as a stop (see Figure 15-17).

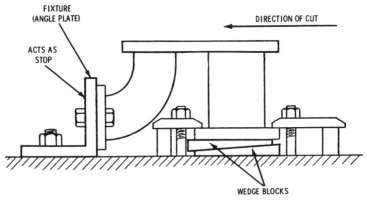

Figure 15-17 Fixture (angle plate) used to hold an irregularly shaped casting. End support with clamps and wedge blocks is also required to prevent springing of the work.

Work Held by a Vise

A vise has only two jaws. One jaw is stationary and the other jaw is adjustable. A chuck has more than two jaws, all of which are adjustable. A vise is often erroneously called a chuck. A machine vise is designed for attachment to a machine table and consists of a base, a table, and both a stationary and an adjustable jaw.

Two general types of vises are the *plain vise* (single and double screw) and *universal vise*. Basic parts of a plain vise are illustrated in Figure 15-18. The base of a machine vise is constructed with a tongue that fits into the table slot, and it has lugs or open holes for bolting to the shaper table with T-bolts. Also, the base has a circular gauge, graduated in degrees, to indicate any angular position. The base has a stationary jaw at one end and ways for movement of the adjustable jaw. It also carries the clamp screw.

When fastening work in the vise, the movable jaw should be brought near its final position for engagement of the workpiece. The jaw hold-down bolts should be tightened tightly. Otherwise, the movable jaw and the work can tilt (Figure 15-19).

Workpieces that project above the vise jaws ordinarily cause no problem in holding. A thin piece of work can be placed on parallel strips to project it above the vise jaws so that the jaws will not interfere with the cutting tool (Figure 15-20). If the workpiece is long and narrow, the vise should be turned so that the cutting tool can be used for longer strokes, rather than shorter strokes.

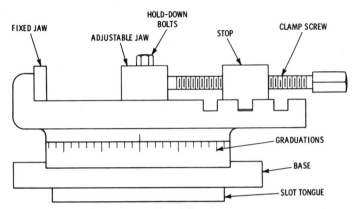

Figure 15-18   Basic parts of a plain vise.

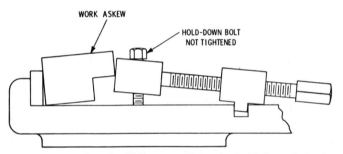

Figure 15-19   Result of failure to tighten the hold-down bolts on the vise.

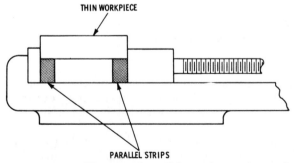

Figure 15-20   The use of parallels to elevate the workpiece above the vise jaws for machining.

Round work that is too small in diameter to be machined while resting on the ways of the vise can be held in horizontal alignment (parallel with the vise table) by means of V-blocks, as shown in Figure 15-21. If an irregularly shaped piece of work having a concave side were clamped in the vise, the jaw on the concave side would not be held properly, and the piece would tilt. This can be avoided by using a round rod to aid in holding the work (see Figure 15-22).

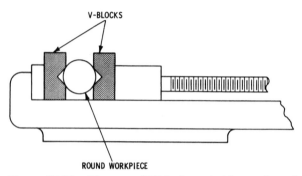

Figure 15-21   The use of V-blocks to hold round stock in the vise.

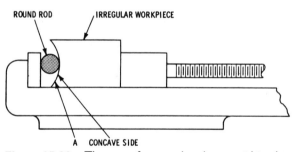

Figure 15-22   The use of a round rod as an aid in clamping irregularly shaped work.

The movable jaw of a universal vise can be swiveled to hold an angular workpiece. A stop can be used to prevent the work from shifting sideways if the side of the piece has too large an angle (Figure 15-23).

When the movable jaw of a vise is forced against the work, the jaw tends to lift upward, tilting the work as the jaw is lifted. This can be avoided by using a round rod. If the jaw should lift upward, the workpiece would remain in its original position (Figure 15-24).

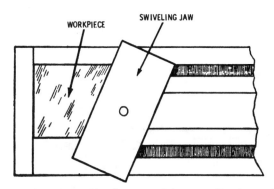

Figure 15-23 Application of the swiveling jaws of a universal vise to hold an angular workpiece.

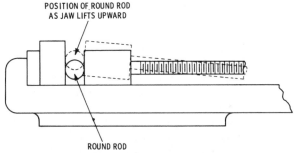

Figure 15-24 Application of a round rod to prevent tilting of the workpiece should the vise jaw lift upward.

Special Fixtures
Special fixtures for a given workpiece can be designed to hold irregularly shaped castings. The fixture must position and support the casting in the correct places to prevent stresses caused by uneven clamping, which tends to spring the work. The special fixtures should be constructed so that they can be attached to the work with a minimum loss of time.

Workpiece Mounted Between Centers
An attachment similar to the lathe headstock and tailstock can be used to mount some kinds of work that cannot be completely machined on a lathe. The headstock spindle does not rotate, except for turning to any desired angular position as determined by an index head (Figure 15-25).

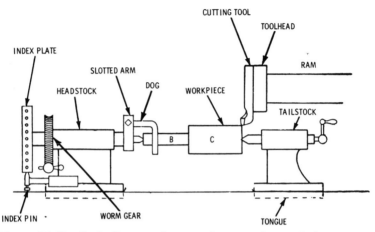

Figure 15-25 Basic diagram of a setup for mounting workpieces between shaper centers.

To set up the work, the same centers that were used on the lathe should be used on the shaper. A tongue on the bottom of the attachment fits the table slots of the shaper; this provides parallelism and alignment. However, the alignment should be checked before the machine operations are begun.

In the setup (see Figure 15-25), the headstock spindle can be rotated by turning the handle of the worm gear. The index plate can be locked in any position by engaging the index pin with the corresponding hole in the circumference of the index plate.

For example, a set of five index plates can have 44, 52, 56, 90, and 96 holes. The index plate that is used must have a number of holes that is evenly divisible by the required number of angular positions that are necessary to machine the piece. For instance, the section B (see Figure 15-25) has been turned in the lathe, and the section C requires six sides that are to be machined in the shaper.

After selecting the index plate with 90 holes (15 × 6 = 90), set the index plate so that the index pin engages the "zero" hole in the plate, and machine one side of the work to the finished dimension. The angular setting for the second side can be obtained by disengaging the index pin, turning the worm gear handle to rotate the index plate 15 holes, and engaging the index pin in the hole.

Adjusting the Work
After the workpiece has been mounted and aligned on the shaper table, the ram should be moved outward until the tool post is over

the work. The tool should be placed in the tool post with as little overhang as possible. The table should be elevated until the workpiece clears the ram by a distance of about one inch. The tool slide should project only a short distance below the bottom of the shaper head (that is, the overhang of the slide should be not more than 1½ inches). If more overhang is necessary, the cutting tool, rather than the slide, should be set for the extra overhang.

The cutting tool should be clamped firmly, making certain that it clears the work. The machine can then be started to note the position and length of stroke.

Stroke Adjustment

The ram should be adjusted to provide the proper length of stroke and to provide the proper position of travel over the workpiece. The stroke adjustment screw (see Figure 15-2) can be turned to produce either a shorter or a longer stroke. A handle is provided on most machines for the square head of the adjusting screw. Unless the operator is familiar with the machine, trial and error can determine the direction of turning the stroke adjustment screw for either a longer or a shorter stroke. The stroke should be long enough for the cutting tool to clear the work by not less than ¼ inch on the forward or cutting stroke. It should clear by ½ inch behind the workpiece on the return stroke. This is so that the automatic feed can function before the cutting tool contacts the work for the forward stroke (Figure 15-26). The automatic feed should function at the end of the return stroke, rather than at the end of the forward or cutting stroke.

If, after proper length of stroke has been obtained, the travel of the cutting tool is not positioned correctly over the work, the position of the ram should be changed. The ram clamp should be loosened and the ram positioning adjustment turned until the cutting tool clears the work by a distance of ¼ inch beyond the workpiece on the forward stroke and ½ inch behind the workpiece on the

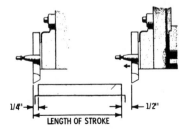

Figure 15-26 Front and back positions of the ram in relation to the workpiece.

return stroke (see Figure 15-26). Then, the ram clamp should be tightened. The operator should remember to adjust the length of stroke before adjusting the position of travel over the workpiece.

Head Adjustment

Movement of the cutting tool and the tool slide is controlled by the vertical feed handle at the top of the head. The handle is attached to

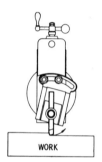

Figure 15-27
Turn the top of the clapper box away from the cutting edge of the tool to permit the tool to swing away from the work on the return stroke.

the downfeed screw inside the slide. A micrometer dial graduated in thousandths of an inch can be used to make accurate adjustments of the slide and cutting tool. Backlash or lost motion between the threads of the downfeed screw and the nut must be removed before the dial can be set accurately.

The toolhead should be positioned at an angle in shaping a dovetail and in certain other setups. The operator should be careful not to run the ram back into the column while the slide is set at an extreme angle, as the slide will strike the column when the ram moves back on the return stroke.

The clapper box should be set so that the top is slanted slightly away from the cutting edge of the tool (Figure 15-27). Then, the tool can swing away from the work on the return stroke of the ram, thus protecting the cutting edge of the tool.

Taking the Cuts

The operator should decide on the number of cuts necessary, depending on the amount of metal to be removed. Usually, one or two roughing cuts and one or two finishing cuts are required, depending on the nature of the work and the finish desired.

The ram should be positioned over the work and the cutting tool adjusted by the vertical feed handle so that it almost touches the work. By means of the hand cross-feed screw, move the table to the right or left (according to the cutting tool in use) until the workpiece is in position for beginning the cut; the tool should be far enough from the work that one or two strokes can be made before the tool begins to cut metal.

Then, set the cutting tool to the proper depth for the first cut by means of the vertical feed handle. The first cut should be deep enough to barely remove the scale, thus truing the surface and enabling the operator to check the setting of the cutting, too. Then, the necessary roughing and finishing cuts can be taken on the workpiece.

Selecting the Proper Feed

In shaping, the feed is the distance the work is moved toward the cutting tool for each forward or cutting stroke of the ram. Either the hand-feed or the automatic feeding mechanism can be used. In setting the automatic feed, the amount of feed will depend on the kind of metal being cut. The time required to complete the work with a given finish is determined by the amount of feed. Both the kind of metal and the type of job must be considered in selecting the feed.

If a softer metal is used, a heavier feed can be used (0.030 to 0.060 inch). A finer feed must be used on harder and tougher metals (0.005 to 0.015 inch). A feed of 0.062 ($\frac{1}{16}$) inch can be used on cast iron.

Of course, the capacity of the shaper must be considered. Heavier feeds can be used on a large heavy-duty machine than on a small and lightly constructed shaper.

Selecting the Proper Cutting Speed

The speed of the shaper is the number of cutting strokes made by the ram during one minute of operation. This is governed by the speed of the bull gear. The number of strokes made by the ram in one minute remains constant for a given speed of the bull gear whether the stroke is long or short. The cutting speed is changed by changing the rate and amount of rocker-arm movement.

The speed of the cutting tool is the average rate of speed attained when the shaper has been adjusted to make a given number of cutting and return strokes of a given length in one minute. The rate of speed, then, is determined by the time or fractional part of a minute required for the cutting stroke and the distance or total length in feet of the cutting stroke.

Cutting speed is determined by the total distance that the tool travels during the cutting strokes made in one minute, and the ratio of cutting-stroke time to return-stroke time. In most shapers, about 1½ times as much time is required for the cutting stroke as for the return stroke, giving a ratio of 3:2. Thus, the cutting stroke requires ⅗ of the time for the cycle (a complete cycle consists of one cutting stroke and one return stroke).

If the length of stroke in inches and the number of strokes per minute are given, their product will give the number of inches of metal cut during one minute of operation. As cutting speed is expressed in feet per minute, it must be multiplied by 12.

Divide the distance (in feet) by ⅗, as the tool cuts during ⅗ of the cycle (distance divided by time equals rate). To shorten the calculation, use the following formula:

Cutting speed = $0.14 \times N \times L$

in which N = strokes per minute, and

L = length of stroke

For example, say you want to find the cutting speed of the tool when the shaper makes 60 strokes per minute and the cutting stroke is 10 inches long.

$$\begin{aligned} \text{Cutting speed} &= 0.14 \times N \times L \\ &= 0.14 \times 60 \times 10 \\ &= 84 \text{ feet per minute} \end{aligned}$$

To determine the number of strokes per minute at which the shaper should be run, with the cutting speed of a metal given in a table, use the following formula:

$$\text{Strokes per minute} = \frac{\text{cutting speed}}{0.14 \times \text{length stroke}}$$

Example

As another example, find the number of strokes per minute at which the shaper should be run to machine a piece of copper having a cutting speed of 80 ft per minute (from table) and with the stroke adjusted to 6 inches.

$$\begin{aligned} \text{Strokes per minute} &= \frac{\text{cutting speed}}{0.14 \times \text{length of stroke}} \\ &= \frac{80}{0.14 \times 6 \text{ inches}} = \frac{80}{0.84} \\ &= 95.238 \text{ strokes per minute} \end{aligned}$$

In most instances, the shaper is operated at a speed that is too slow. Therefore, the beginner should determine the speed at which the shaper should be operated and work as near the speed as possible. Cutting speeds for shaping the common materials are provided in Table 15-1. The length of stroke can be obtained from the machine.

Table 15-1 Shaper Cutting Speeds

| | Cutting Speed (ft/min.) | | Lubricant | |
Materials	H.S. Tools	C.S. Tools	Roughing	Finishing
Aluminum	125	50	Dry	Kerosene
Brass	Highest possible	35	Dry	Dry
Cast iron (hard)	35	20	Dry	Soda water
Cast iron (soft & medium)	60	30	Dry	Dry
Copper	80	40	Dry	Dry
Steel (hard)	35	15	Cutting compound	Lard oil
Steel (soft)	Highest possible	35	Cutting Compound	Soda water

Summary

The shaper is a metal-removing machine with cutting tools that move in a horizontal plane in a reciprocating motion. The size of a shaper is generally determined by the size of the largest cube that can be machined on it.

There are two types of shapers—the crank and the geared type. The crank type is the most commonly used and is made in either a standard or universal type. The shaper was intended originally to produce only plane surfaces (such as angular, horizontal, or vertical), but it can be used to produce either concave or convex surfaces as well.

Basic parts of the shaper are base, crossrail, saddle, table, ram, and bull gear. The base naturally supports the machine and all working parts. The crossrail contains the table elevating and traversing mechanism and is attached to the base by means of gib plates and bolts.

Many factors can affect the efficiency of operation of the shaper. Some of these factors are selection of the cutting tool, holding the work, adjustment of the work, adjustment of the stroke, and the selection of the proper feed and cutting speed.

Cutting tools commonly used on the shaper are right-hand roughing, round nosed, squared nosed, and gooseneck. There is also the right-hand down-cutting tool and the left-hand side-facing tool. One of the special cutting tools used on shapers is for cutting T-slots.

Review Questions

1. Name the two types of shapers.
2. Of the two types of shapers, which is more commonly used?
3. What is the basic operation of the shaper?
4. What are the basic parts of the shaper?
5. What is the purpose of the crossrail?
6. What is a bull gear?
7. What is a ram?
8. Name five of the tools most commonly used on the shaper.
9. The operator should remember to adjust the length of the stroke before adjusting the _____ of travel over the workpiece.
10. The speed of the shaper is the number of cutting _____ made by the ram during one minute of operation.

Chapter 16

Planers

The planer is a machine tool used to machine flat or plane surfaces on work that is fastened to a reciprocating table. The surfaces can be horizontal, vertical, or angular. The planer can also be used to form irregular or curved surfaces.

The planer differs from the shaper in that the worktable moves back and forth with a reciprocating motion while the cutting tool is held stationary, except for transverse movement. The planer is used for the same purpose as the shaper, but it can handle much larger work, and heavier cuts can be taken. The planer is used on work that is too large or otherwise impossible to machine on a shaper. Some typical examples of planer work are machine bases, machine covers, and column supports.

Types of Planers

One of the more common types of planers is the *double-column planer*. Two columns support the crossrail and house the elevating screws and controls for the machine.

The *open-side planer* has only one column or housing to support the crossrail and toolheads. One advantage of this type of planer is that workpieces of irregular shape can be handled with the workpiece extending outward over the side of the table. The working parts are essentially the same as on the double-column planer.

The rated size or capacity of a planer as given by the manufacturer refers to the width, height, and length of the largest workpiece that can be machined. For example, a "24 × 24 × 48 planer" indicates that the planer can machine a piece of work 24 inches wide, 24 inches high, and 48 inches in length.

Basic Construction

The most important parts of a planer are the bed, columns, crossrail, table, toolhead, and table drive (Figure 16-1 and Figure 16-2). The working parts are essentially the same on both double-column and open-side planers.

Bed

The *bed* is an extremely heavy, rigid casting. The table slides in accurately finished ways on the bed. The bed also supports the columns and all moving parts of the machine.

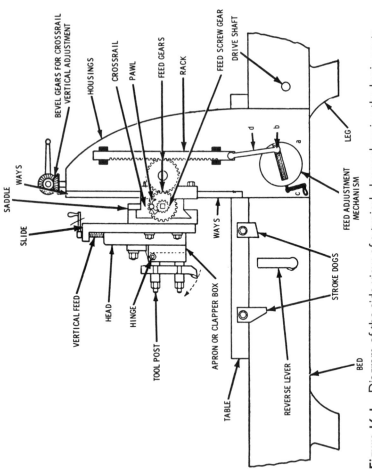

Figure 16-1 Diagram of the side view of a typical planer, showing the basic parts.

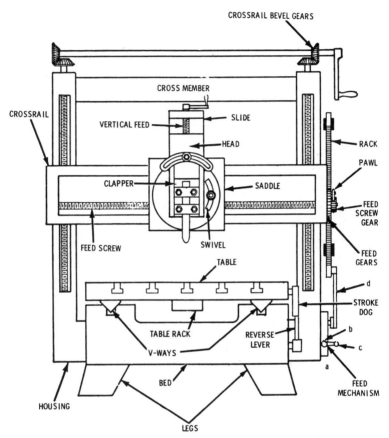

Figure 16-2 Diagram of the front view of a typical two-column planer, showing the basic parts.

Columns

On a double-column planer, two *columns* rise vertically at the sides of the machine. They support the crossrail and house the elevating screws and controls for the machine. The open-side planer has only one column.

Crossrail

The *crossrail* carries the saddle and toolhead. It is a rigid casting and provides guides for transverse travel of the saddle. The crossrail also contains the feed rod and screw for controlling the movement of the cutting tool.

Vertical screws located in each of the columns provide a means of support and a means of adjusting the crossrail vertically. It is of utmost importance that the crossrail, when clamped, be parallel to the table, because the accuracy of the surfaces produced is dependent on the accuracy at which the cutting tool is moved (Figure 16-3). This parallelism can be determined by placing an indicator in the tool post with the point of the indicator touching the table. Then, move the toolhead crosswise of the machine, observing at the same time any variation of the indicator needle.

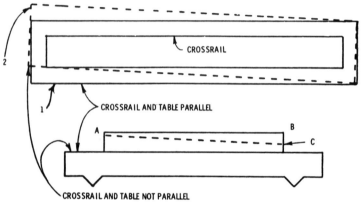

Figure 16-3 Diagram showing the lack of parallelism of the crossrail and table.

Table

The planer *table* is a precision-machined plate that travels on ways of the bed. It presents a broad surface for mounting the workpiece. The table is provided with T-slots for bolting workpieces and accurately reamed holes for locating stops. Both T-slots and holes should be kept free from nicks and burrs. The shanks of the stops should never be forced into the holes by driving them with a hammer; this action upsets the surface of the table and destroys its accuracy.

Toolhead

The *toolhead* of a planer is similar to that of a shaper both in construction and in operation. A feed screw is provided to move the toolhead with respect to the work. The toolhead can be swiveled for taking angular cuts, and it can be set over in either direction to provide tool clearance when taking vertical or angular cuts on the planer.

Table Drive

The *table drive* of a planer can be obtained by one of three methods:

- Rack and spur gears
- Spiral rack and worm
- Crank

Figure 16-4 shows a basic diagram illustrating the rack-and-spur-gear method with belt-pulley drive.

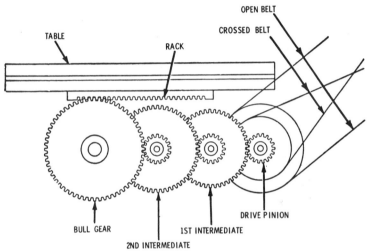

Figure 16-4   Diagram of rack-and-spur-gear table drive mechanism.

When the drive mechanism is made up entirely of gears, a large gear known as a *bull gear* is connected to a rack on the bottom side of the table and to an electric motor by a series of gears. The quick return of the planer table is accomplished by adjustable stops on the side of the table, which, at the end of each cutting stroke, come in contact with a lever that engages a high-speed gear in the driving train of gears.

In the belt-driven type of planer, two belts, one open and one crossed, operate on loose and fixed pulleys. The table travel stops come in contact with a shifter lever at the end of each stroke. As it makes contact at the end of the cutting stroke, the lever throws the belt off the fixed pulley onto the loose pulley. At the same time, it throws the crossed belt from a loose pulley to a fixed pulley.

Because the crossed belt is running faster than the open belt, the table moves faster on the return stroke. The operator can shift the lever by hand to run the belts on loose pulleys, thus stopping the movement of the table without stopping the whole machine. The lever can be locked in neutral position to prevent accidental starting of the machine.

The open- and crossed-belt drive mechanism permits operation of the gear train in such a manner that the table will travel slowly on the cutting stroke, reverse, and travel faster on the return stroke. Four pulleys (two larger pulleys and two smaller pulleys) are required (Figure 16-5). One of the larger pulleys and one of the smaller pulleys are keyed to the drive pinion shaft and are called *tight pulleys* to distinguish them from the other two pulleys, which turn freely on the shaft and are called *loose pulleys*.

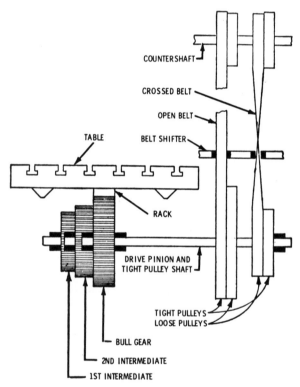

Figure 16-5 Basic diagram of open- and crossed-belt drive of a planer table drive mechanism.

The larger tight pulley is used for the slower forward speed (or cutting-stroke drive), and the smaller tight pulley is used for the quicker return stroke. Both belts are driven by wide-faced pulleys of the same diameter placed on the countershaft. The open belt on the larger tight pulley drives the machine at a slower speed than the crossed belt on the smaller tight pulley. In actual operation, both belts run continually and can be shifted back and forth by the belt shifter, which is linked to the reverse lever.

Planer Tools

Cutting tools used on the planer are similar to those used on the lathe (Figure 16-6). However, the planer is used to machine flat surfaces, while circular surfaces are machined in lathe operations. The chief difference in cutting principles is that the cutting tool on a lathe tends to spring away from the work when it is set at exact center height, but the cutting tool on a planer tends to dig into the work if its cutting edge is set in advance of the plane of support (Figure 16-7). This can be avoided if the tool is forged so that the cutting edge is behind the plane of support. The characteristics of the workpiece determine the number and variety of cutting tools required.

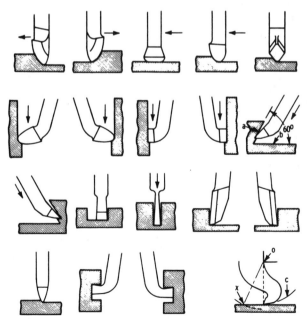

Figure 16-6   Kinds of cutting tools commonly used on planers.

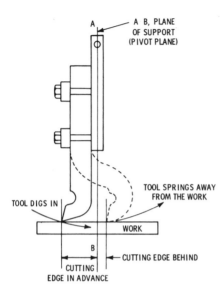

Figure 16-7 The reason for placing the cutting edge of the planer tool behind the plane of support is to prevent "digging in" of the tool.

Planing Operations

A variety of clamps are available for holding the workpiece on the planer table. Clamping involves the use of such items as bolts, studs, washers, shims, nuts, step blocks, toe dogs, stops, strap clamps, and C-clamps.

Planing Horizontal Surfaces

A typical example of work that can be machined horizontally is the flanged cast-iron cover (Figure 16-8). This kind of casting is not difficult to clamp to the table. Mount the cast-iron cover with its flanged side downward on the planer table, adjust the crossrail to the correct height, and set up the clamp screws.

The roughing tool should be placed in the toolholder with the cutting tool perpendicular to the work. The tool should be placed against the two clamping bolts on the side so that lateral thrust cannot cause the tool to shift. Tighten the cutting tool bolts so that the clamps are parallel with the clapper—not tilted.

Position and set the stroke for machining surface A (see Figure 16-8). The depth of the roughing cut depends on the amount of metal to be removed. A deep cut cannot be taken on hard metals because the metal will break and leave a ragged edge at the end of the stroke. Check for this condition at the end of the first stroke. If the metal tears, set the tool for a lighter cut and take two roughing

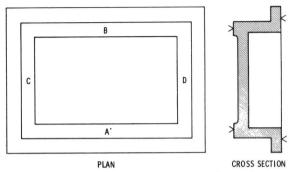

PLAN   CROSS SECTION

Figure 16-8 Diagram (left) and sectional view (right) of a flanged cover as a typical workpiece for showing the planing of horizontal surfaces.

cuts if necessary. Machine until the cutting tool reaches the inside edge of surface *A*.

Position and set the stroke to machine surface *D* to the inside edge of surface *B*. Similarly, after machining surface *D*, reset and position the stroke for surface *B*. Resetting can be avoided on small workpieces by planing the entire surface at one setting of the cutting tool.

The metal should be roughed until about 0.015 inch of metal is left for finishing. Prepare the work for finishing by breaking the front edge of the surface (the edge at which the tool begins to cut) with a coarse double-cut file. This is done to keep the scale on the outside surface of the casting from destroying the fine cutting edge of the finishing tool.

In taking the finishing cut, extreme care should be taken to finish to the dimension of the blueprint. Remember that if the first cut does not remove enough metal, another light cut can be taken. If the first cut removes too much metal, the work can be ruined.

When the casting is turned over to machine the second side, the casting is automatically leveled, because the machined surface will be in contact with the table surface. Therefore, the two sides should be parallel after both sides have been machined. Of course, different clamping methods are necessary to hold the work for machining the second side.

Thin castings should be rough-planed on all sides first and then finished. Certain conflicting stresses are set up in the piece when it is cast, because the outer surface becomes chilled first. This places a stress on the scale, or skin, of the casting. Removal of this scale allows the metal to adjust itself to these internal stresses, and the surface will not be true when it is planed.

Planing at an Angle

Planing must be performed at a given angle for dovetails, V-shaped grooves, and so on. The toolhead assembly must be pivoted with respect to the saddle so that it can swing around a center axis.

Graduations on the circular part indicate the angular settings in degrees. However, it cannot be assumed that the correct angle is indicated when the head is set at a given angle on the scale (especially on old and worn planers) because more or less lost motion can be present. Therefore, after setting by scale, the setting should be checked as shown in Figure 16-9.

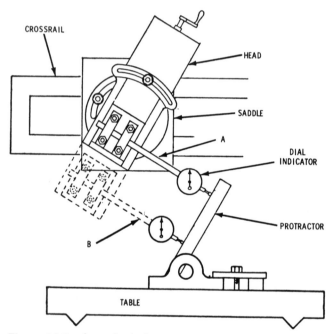

Figure 16-9   A method of using a protractor and dial indicator to check the angular setting of the slide head.

Mount a dial indicator in the toolholder, clamp a protractor to the table, and set the head at the desired angle (see Figure 16-9). Adjust the head until the dial indicator contacts the protractor (position *A*). Feed downward until the indicator reaches position *B*, and note the reading. If the two readings are the same, the setting is correct.

The planer table V-block (Figure 16-10) can be used as an example for planing at an angle. As shown in the end view, the sides of the V-block are at 45° from the vertical axis. Set the toolhead at 45° on the scale, as shown in Figure 16-11. Clamp the head in position by tightening the bolt *A*. Set the tool block at an angle less than 45°, and clamp in position by tightening the bolt *B*. The tool block is set at less than 45° to prevent the cutting tool from dragging over the planed surface on the return stroke (Figure 16-12). By setting the tool block in the angular position, the cutting edge of the tool swings forward in the plane *YY* at right angles to the tool swing axis *XX* from the work. The angular position of the cutting tool does not affect the direction of tool travel.

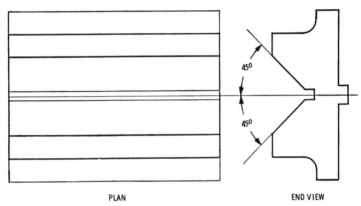

PLAN   ENDVIEW

Figure 16-10 Diagram (left) of a planer table V-block with the end view (right) as an illustration of planing at an angle.

The saddle should be moved into position for the first roughing cut (see Figure 16-11) on the V-block (it is assumed that the bottom side, with its tongue, has been machined). The first cut should be taken at the top. Start the cut with the hand-feed, moving downward. Then, engage the slide-feed, causing the slide and tool to feed downward for each stroke—the saddle remaining in the same position on the crossrail.

After completing the roughing and finishing cuts on one side, do not change the angular setting of the slide for the outer side of the vee. Reverse the V-block (that is, turn it 180°) in the table slot, and machine the other side of the vee. This is done because it is impossible to reset the slide to exactly the same angle. Moreover, reversing the work not only ensures machining at the same angle but also

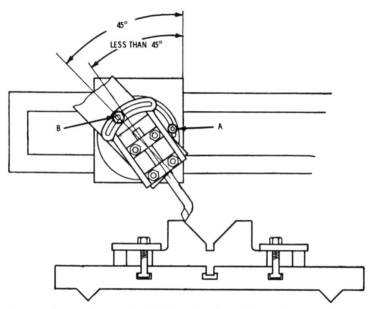

Figure 16-11 Setting the slide head and tool block for planing the V-block.

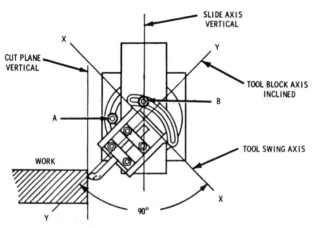

Figure 16-12 Diagram showing the reason for turning the top of the tool block "away from" the surface to be planed when planing either vertical or inclined surfaces.

ensures that the vee will be in alignment with the center axis of the tongue.

Planing Curved Surfaces

A fixture consisting of a radius arm pivoted on a bracket can be used to plane a concave surface (Figure 16-13). The feed screw of the slide is removed, and the slide is fastened to the radius arm.

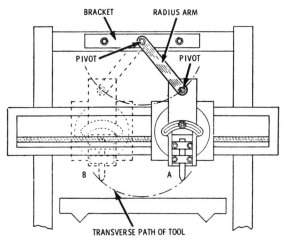

Figure 16-13   Diagram of a fixture that can be used to plane a circular surface.

In planing, the cross-feed causes the saddle to traverse the cross-rail as the tool, which is guided by the radius arm, planes a curved surface. The height of the cutting edge of the tool, as the saddle traverses the crossrail, is determined by the angular position of the radius arm (positions *A* and *B* in Figure 16-13). The locus of all these positions is indicated by the arc.

Planing a Helix (Work on Centers)

A helix that has a long pitch cannot be cut on some milling machines, but it can be produced on a planer by mounting the work between planer centers and using the fixture, as shown in Figure 16-14. The fixture consists of a weighted clamp bar and an inclined bar. The upper end *A* of the inclined bar is attached to the housing, and the lower end *B* of the inclined bar is attached to the planer bed.

The pitch of the helix depends on the inclination of the bar. The cutting tool is formed to produce the desired helical groove. As the

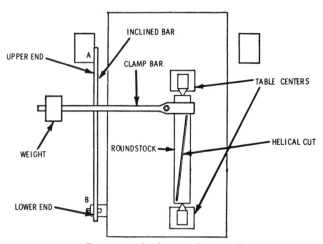

Figure 16-14   Diagram of a fixture that can be used to cut a helix.

table moves with the clamp bar near the lower end of the inclined bar, the clamp bar (being guided as to angular position) gradually rises as the work turns clockwise through a small arc. Thus, the cutting tool is caused to cut a helix, as shown in Figure 16-14.

Summary

The planer is a machine that is used on flat or plane surfaces that are fastened to a reciprocating table. The planer differs from the shaper in that the worktable moves back and forth with a reciprocating motion, while the cutting tool is held stationary. The planer is used for the same purpose as the shaper, except the planer can handle much larger work and heavier cuts can be taken. The planer is used on work that is too large or otherwise impossible to machine on a shaper.

The most common type of planer is the double-column machine. These columns support the crossrail and house the elevating screws and controls. Another type of planer is the open-side planer, which has only one column or housing to support the crossrail and toolhead. One advantage of this type of planer is that workpieces of irregular shape can be handled with the workpiece extending outward over the side of the table.

Cutting tools used on the planer are similar to those used on the lathe. The chief difference in the cutting principle is that the cutting tool on a lathe tends to spring away from the work, while it tends to dig into the work on a planer. The planer is used to machine flat surfaces, while the lathe machines cylindrical surfaces.

The most important parts of a planer are the bed, columns, crossrail, table, toolhead, and table drive. The working parts are essentially the same on both double-column and open-side planers.

The planer table is a precision-machined plate that travels on ways of the bed. It presents a broad surface for mounting the workpiece.

A variety of clamps are available for holding the workpiece on the planer table. Clamping involves the use of such items as bolts, studs, washers, shims, nuts, step blocks, toe dogs, stops, strap clamps, and C-clamps. A helix that has a long pitch cannot be cut on some milling machines, but it can be produced on a planer by mounting the work between planer centers and using a fixture that consists of a weighted clamp bar and an inclined bar.

Review Questions

1. What is the most common type of planer used?
2. What are the basic operation differences between the planer and the shaper?
3. What are the two types of planers manufactured?
4. What is the advantage of the planer over the shaper?
5. What is the purpose of the crossrail?
6. Describe the table on the planer.
7. What is a toolhead similar to?
8. Name three methods of driving the planer table.
9. Cutting tools on the planer are similar to those used on the _____.
10. A fixture consisting of a radius arm pivoted on a bracket can be used to plane a _____ surface.

Chapter 17

Slotters

A *slotter* (or *slotting machine*) is, in many respects, a heavy-duty vertical shaper. The chief difference between the slotter and the shaper is that the ram moves in a vertical direction on the slotter.

The slotter can be used for a variety of work other than slotting. Although its original purpose has been changed, the machine is still referred to as a slotter. In addition to slotting work, several other kinds of workpieces can be machined more advantageously by a tool that cuts in a vertical direction. Regular and irregular surfaces (both internal and external) can be machined on the slotter, and it is especially adaptable for handling large and heavy pieces of work that cannot be handled easily on other machines.

Basic Construction

A diagram of a typical crank-driven slotter is shown in Figure 17-1. The *ram* carries the cutting tool and has a reciprocating motion. It is similar to the ram of a shaper, but it is more massive and moves vertically, or at right angles to the table, instead of having the horizontal motion of a shaper. A connecting rod links the crank with the ram and changes the rotary motion of the crank to reciprocating motion, which causes the ram to move up and down. A stroke-adjusting screw can be turned to move the crank either toward or away from the crank disk center, which either shortens or lengthens the stroke of the ram.

The *table* of a slotter can be moved either transversely or longitudinally, or it can be rotated about its center. The hand-feed mechanism consists of three screw rods, operated by a crank handle, which can be attached to the squared ends of the feed rods. The positions for the crank handle are indicated by the letters *A*, *B*, and *C* in Figure 17-1, and they are for operating the transverse, longitudinal, and circular feeds, respectively.

The power-feed mechanism is, of course, more complicated than the hand-feed mechanism. It consists of a reciprocating oscillating unit and a transmission gearing unit. The feed mechanisms of various machines differ in detail from the basic diagram (see Figure 17-1), but the basic principles of operation are similar for all machines.

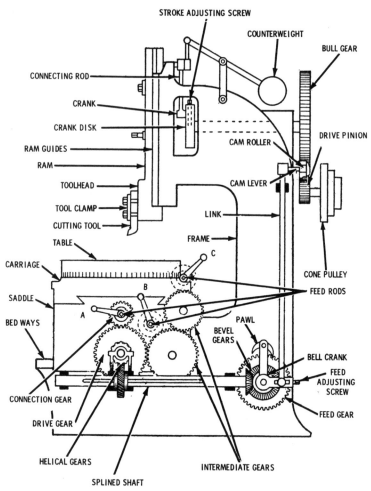

Figure 17-1 Diagram of the basic parts layout of a crank-driven slotter.

Slotter Operations

Workpieces that are to be machined on the bottom side cannot be mounted directly on the table surface but should be mounted on parallel strips to give clearance for the cutting tool (Figure 17-2). The height *A* of the parallel should be a large enough distance to provide work clearance. Since workpieces that are to be machined on the bottom side cannot be mounted directly on the table surface

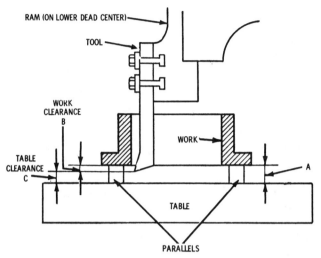

Figure 17-2 Proper clearances for a cutting tool, work and table.

they should be mounted on *B*, and there should be ample table clearance *C* for the cutting tool at the end of its cutting stroke. A broken cutting tool and serious damage to the machine can result from insufficient table clearance.

Clamping the Work

The same clamping methods used for shapers and planers can be used on slotters. Although the ram has a vertical stroke instead of a horizontal stroke (as in the shaper and planer), the cutting stroke of the tool exerts a side pressure on the workpiece. This can be counteracted by proper placement of stops to prevent any possibility of the movement of the work. Circular workpieces should be centered with respect to the center of the table.

Adjusting Stroke Position and Length of Stroke

Place the workpiece on parallels in its approximate position to barely clear the cutting edge of the tool. Before placing the cutting tool in the ram head, place the ram on bottom dead center. Then, adjust the cutting tool to travel ¼ inch below the bottom side of the work. The tool should travel above the topside of the work an ample distance for the power feed to operate before the cutting tool begins the next cut.

On some machines, the stroke-adjusting screw is inconveniently reached when the crank is "on center." If this occurs, the crank disc

344 Chapter 17

can be given a quarter-turn so that the adjusting screw can be turned. The stroke can then be estimated "by eye" and readjusted until a stroke of proper length is obtained. A stroke indicator with a graduated scale is available on some machines so that a stroke of proper length can be obtained with one setting, thus avoiding the trial-and-error method.

Alignment of the Work
After the work to be machined has been carefully laid out with clear and distinct scriber lines, place a scriber in the ram head so that its point barely clears the work. Adjust the table position with the transverse hand-feed (*A* in Figure 17-1) so that the scriber is opposite one end of the line scribed on the work. Then, bring the scriber line directly below the scriber tool by turning the longitudinal hand-feed *B*.

Check the setting of the work with the transverse hand-feed. As the work moves, the scribed line on the work should follow the scriber mounted in the ram head if the work has been aligned properly. Otherwise, adjustment of the circular position is necessary. To adjust the circular position of the table, turn the circular feed rod handle *C*, and check as before. Repeat angular adjustment until the work is properly aligned transversely.

Straight, or Flat, Slotter Work
A rectangular open-box-like piece of work can be used to illustrate straight (or flat) machining on the slotter (Figure 17-3). The four sides of the box are to be machined externally to the scribed lines.

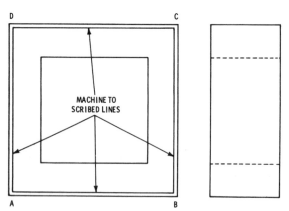

Figure 17-3   Front view (left) and side view (right) of an open-box-like casting used to illustrate machining on the slotter. Note the scribed layout lines for indicating finishing limits.

Select parallels, place the work on the parallels, and clamp securely to the table. If the stroke position and length are adjusted properly, the cutting tool should be in the position shown in Figure 17-4 when the ram is on bottom dead center.

The cut should be started at one end for the side *AB* (see Figure 17-4) and the entire side machined by vertical downward strokes combined with the transverse feed. Similarly, the opposite side *DC* should be machined without disturbing the mounting of the workpiece on the table by raising the ram position to top dead center and turning the cutting tool 180°. That is, the cutting tool should be in position 1 for machining *AB* and in position 2 for machining the opposite side *DC* (see Figure 17-4). To machine the second side, the tool can be brought into position by means of the longitudinal feed screw and fed by the transverse feed.

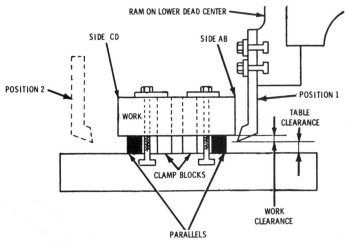

Figure 17-4   Setup for machining the open-box-like casting. The dotted lines (tool position 2) indicate the position of the cutting tool after reversal.

For the other two sides (*BC* and *AD*), turn the table 90° by means of the circular feed rod (*C* in Figure 17-1), and proceed to machine the sides as before. On some slotters, the ram head (which carries the cutting tool) can be swiveled to any angular position. Four graduations are provided (90° apart). Thus, all four sides of the work can be machined by swiveling the tool, making it unnecessary to reverse the cutting tool or to change the angular setting of the table. For precision machining, the scribed lines merely serve as a guide. The final finishing cut should be made to precise caliper measurements.

Circular Slotter Work

The slotter is well suited for machining cylindrical surfaces (Figure 17-5). The work must be properly aligned (that is, centered with the center of the table). In other words, the work must be positioned on the table so that the axis of the cylindrical surface to be machined coincides with the axis of rotation of the table. The circular feed screw (C in Figure 17-1) rotates the table to produce a small arc after each cutting stroke of the tool.

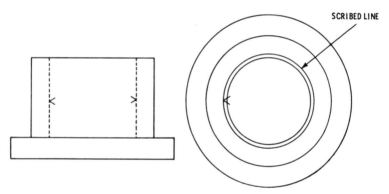

Figure 17-5 Front view (left) and top view (right) of a flanged casting that is typical of circular slotter work.

The center hole of the table is used as a guide for centering small workpieces. Some machines have concentric circles cut in the table for centering large workpieces. Figure 17-6 shows the setup for machining a simple flanged cylinder (see Figure 17-5).

The workpiece should be centered (with precision) concentric to the center of the table. If the table does not have a center hole, a center hole should be bored, and a center fitting should be machined to fit the hole (Figure 17-7). A center fitting should be included as an accessory for all slotters. The fitting should be machined accurately and stored with the same care that should be given calipers and other delicate measuring equipment.

To center the workpiece, clamp a scriber in the ram head, so that its point will barely clear the top of the center fitting. Shift the position of the table by means of the transverse and longitudinal feeds until the point of the scriber is directly over the center of the fitting. After centering the scriber tool, do not change the longitudinal setting of the table.

Raise the ram to upper center to place the scriber out of the way. Place the workpiece on parallels on the table in an approximate

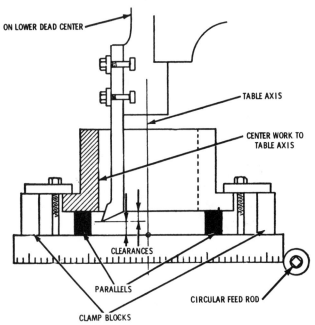

Figure 17-6 Setup for machining the flanged cylindrical casting. Note the table axis on which the work should be centered.

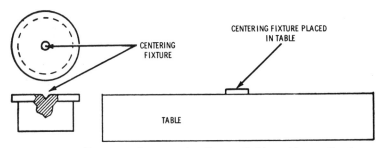

Figure 17-7 Diagram of a centering fixture (left), and its application as a table accessory (right) for the slotter.

concentric position "by eye," and true up with hermaphrodite calipers (Figure 17-8). Set the calipers to one-half the bore diameter, making allowance for thickness of the metal to be removed.

In truing up with the hermaphrodite calipers, caliper two radii on the same diameter, shifting the work along this diameter until

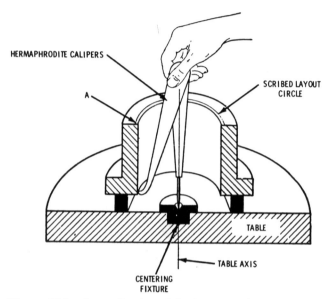

Figure 17-8 An application of the centering fixture and the hermaphrodite calipers for centering circular work. Note the scribed layout circle that can be used as a reference guide for the final centering check.

centered. Repeat on a diameter at 90° (located by eye). Clamp the work lightly, remembering not to change the longitudinal setting of the table. Move the table over with the transverse feed, and lower the ram until the point of the scriber previously clamped in the ram head is directly over and barely clears the point (A in Figure 17-8) on the scribed layout circle.

Remove the scriber, and clamp the roughing tool in position (see Figure 17-2). Raise the arm to bring the cutting tool slightly above the top surface of the work. Shift the table with the transverse feed to bring the tool to proper axial position (A in Figure 17-9). Clamp the carriage to the saddle to prevent any transverse movement, and start the machine.

After two or three strokes of the cutting tool, check to make certain that the stroke and cutting tool are adjusted properly. Then, feed the cutting tool into the work to the proper depth for a roughing cut by means of the longitudinal feed, and lock the saddle to prevent any change in the setting of the longitudinal feed. Using the

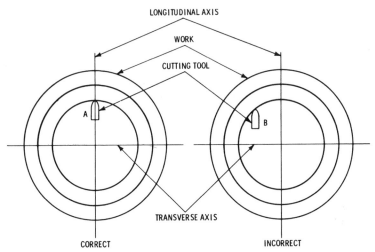

Figure 17-9   The correct and incorrect positions for the slotter cutting tool.

circular feed, rough cut to near the layout circle, leaving only enough metal for the finishing cut or cuts.

Slotter Cutting Tools

As the cutting motion of a slotter is vertical (rather than horizontal), slotter cutting tools differ from shaper and planer tools. The cutting edge is formed on the end of the cutting tool. This is so that it is placed under compression and will cut when pushed endwise. The cutting face that turns the shaving is on the end of the tool (Figure 17-10).

The horizontal axis (*AB* in Figure 17-10) is parallel with the table and perpendicular to the vertical axis (*CD* in Figure 17-10). The clearance angle (*COG* in Figure 17-10) should be 4° or 5°, and the rake angle (*BOF* in Figure 17-10) should be 10° or 12°. When the cutting tool is clamped to the end of the ram in a horizontal position, the clearance and rake angles are measured in the same manner as for planer tools.

A typical set of slotter cutting tools includes the following:

- Roughing
- Finishing
- Right-hand

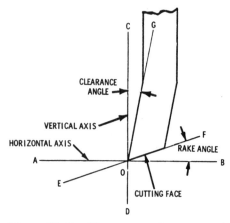

Figure 17-10 The correct clearance and rake angles for the slotter cutting tool.

- Left-hand
- Keyway
- Scriber

The pointed scriber is a much-used tool for cylindrical work setups. In addition to these tools, several special tools are designed for certain kinds of work. For example, a heavy slotter bar with an inserted cutting tool is better adapted for external cuts than tools that are forged in a single piece. Properly held in a round bar that can be turned around, the inserted cutting tool can be set at any angle (Figure 17-11). This adapts the tool for machining slots or keyways, using a broad-nosed cutter of desired shape and width. For cutting square holes, the bar can be turned to reach the corners. Cutting tools inserted in a bar should not be given too much rake because the tool drags over the work on the upward stroke.

Feeds and Speeds

Longitudinal, transverse, and circular feeds are provided on the slotter as both hand- and power-feeds. Rapid power traverse is provided on most slotters, particularly the larger sizes of slotters. They are constructed so that the feed and rapid power traverse are interlocking. Therefore, it is impossible to engage both at the same time. Approximate cutting speeds for various metals are provided in Table 17-1.

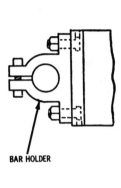

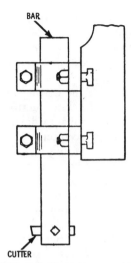

Figure 17-11 Detail of a holder for a round bar (left) and a round bar and with inserted cutting tool (right). Note that the cutting tool is in a horizontal, rather than a vertical, position.

Table 17-1 Slotter Cutting Speeds (Feet per Minute)

	Roughing		Finishing	
Material	Carbon Steel Tools	High-Carbon Steel Tools	Carbon Steel Tools	High-Carbon Steel Tools
Aluminum	50	125	40	60
Babbitt	50	Highest	30	80
Brass	35	Highest	30	80
Cast iron (hard)	35	20	10	20
Cast iron (soft & med.)	30	60	15	45
Copper	40	80	20	30
Monel metal	40	80	20	30
Steel (hard)	15	35	10	25
Steel (soft)	35	Highest	25	50

Summary

A slotter, in many respects, is a heavy-duty vertical shaper. The slotter can be used for a variety of work other than slotting. In addition to slotting work, several other kinds of pieces can be machined more advantageously by a tool that cuts in a vertical direction. Regular and irregular surfaces, both internal and external, can be machined. The slotter is especially adaptable for handling large and heavy pieces of work that cannot be handled easily on other machines.

Because the cutting motion of a slotter is vertical (rather than horizontal), slotter cutting tools differ from the shaper and planer tools. The cutting edge is formed on the end of the cutting tool so that it is placed under compression and will cut when pushed end-wise. The cutting face that turns the shaving is on the end of the tool. When the cutting tool is clamped to the end of the ram in a horizontal position, the clearance and rake angles are measured in the same manner as for planer tools.

A typical set of slotter cutting tools includes the roughing, finishing, right-hand, left-hand, keyway, and scriber. The scriber is often used for cylindrical work setups. In addition to these tools, several special tools are designed for certain kinds of work. Heavy slotter bars with an inserted cutting tool are better adapted for external cuts than tools that are forged in a single piece.

Cutting tools inserted in a bar should not be given too much rake because the tool drags over the work on the upward stroke.

Longitudinal, transverse, and circular feeds are provided on the slotter as both hand- and power-feeds. Rapid power traverse is provided on most slotters, particularly the larger sizes of slotters. They are constructed so that the feed and rapid power traverse are interlocking. Therefore, it is impossible to engage both at the same time.

Review Questions

1. What is a slotter used for?
2. How does a slotter differ from a shaper?
3. What are the basic parts of the slotter?
4. How is the work clamped for slotting?
5. Why is the slotter well suited for machining cylindrical surfaces?
6. Where are the hermaphrodite calipers used?
7. List five of the tools in a typical set of slotter cutting tools.
8. The cutting motion of a slotter is _____.
9. Longitudinal, transverse, and _____ feeds are provided on the slotter as both hand- and power-feeds.
10. What are the rough cutting speeds for brass with H.S. tools?

Chapter 18
Abrasive Metal Finishing Machines

There are two major types of grinding: *nonprecision* and *precision* grinding. Nonprecision grinding is also called *offhand grinding*. Metal is removed by this method when there is no great need for accuracy. Pedestal or bench grinders can be used for this type of grinding (Figure 18-1 and Figure 18-2). These grinders are also used for rough grinding and sharpening tools.

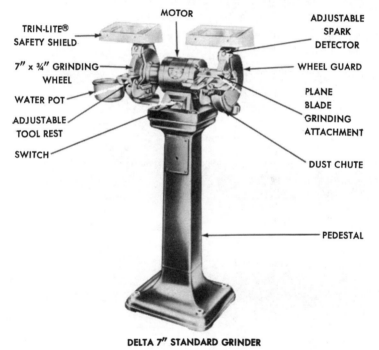

MOTOR

ADJUSTABLE
SPARK
DETECTOR

TRIN-LITE®
SAFETY SHIELD

7″ x ¾″ GRINDING
WHEEL

WHEEL GUARD

WATER POT

PLANE
BLADE
GRINDING
ATTACHMENT

ADJUSTABLE
TOOL REST

SWITCH

DUST CHUTE

PEDESTAL

DELTA 7″ STANDARD GRINDER

Figure 18-1   Pedestal grinder. *(Courtesy Rockwell Manufacturing)*

In precision grinding, metal can be removed with great accuracy. There are a number of different precision grinding machines available that can grind metal parts to different shapes and sizes with very accurate dimensions (Figure 18-3 and Figure 18-4).

353

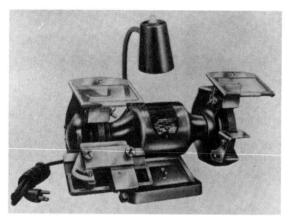

Figure 18-2 A 6-inch edge tool bench grinder.
(Courtesy Rockwell Manufacturing)

Figure 18-3 Form-relief grinder. A single machine for complete form-relief grinding and precision sharpening of a large variety of cutting tools. *(Courtesy Giddings and Lewis Machine Tool Co.)*

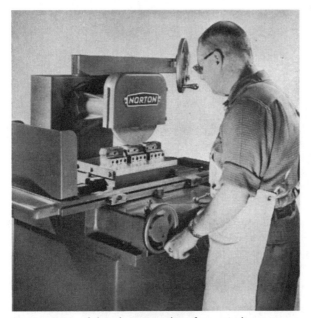

Figure 18-4 A hand-operated surface grinder. *(Courtesy Norton)*

The grinding wheels used in both nonprecision and precision grinding are made from aluminum oxide or silicon carbide. These grinding wheels come in many different shapes, faces, and sizes. Manufacturers of grinding wheels have standardized these shapes and faces (Figure 18-5 and Figure 18-6). The wheels are made in a wide range of sizes for most grinding jobs. A wide variety of

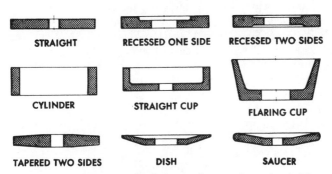

STRAIGHT   RECESSED ONE SIDE   RECESSED TWO SIDES

CYLINDER   STRAIGHT CUP   FLARING CUP

TAPERED TWO SIDES   DISH   SAUCER

Figure 18-5 Nine standard shapes for grinding wheels. These shapes will perform many jobs. *(Courtesy Cincinnati Milacron Co.)*

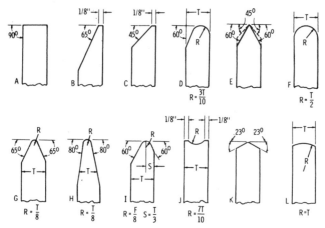

Figure 18-6 Standardized grinding wheel faces. These faces can be modified by dressing to suit the needs of the user. *(Courtesy Cincinnati Milacron Co.)*

Figure 18-7 Mounted points are tiny grinding wheels permanently mounted on small-diameter shanks. They may be as small as $1/16$ inch in diameter. *(Courtesy Cincinnati Milacron Co.)*

mounted grinding wheels are also made for use with small grinders for either offhand or precision grinding on dies (Figure 18-7).

Grinding wheels should be properly mounted on grinders because they are operated at high speeds (Figure 18-8). When the wheels become loaded with particles of material from a grinding operation, they must be dressed. Dressing restores the grinding wheel to its original shape with a clean cutting face.

The three types of grinding wheel dressers in use for precision grinding are the mechanical dresser, the abrasive wheel, and the diamond tool. The diamond tool is the most commonly used wheel dresser on precision grinders (Figure 18-9).

Grinding wheels vary by the following:

- *Type of abrasive*—Aluminum oxide or silicon carbide.

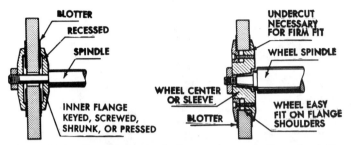

Figure 18-8 Correct method of mounting a grinding wheel. Never omit the blotting paper washers. *(Courtesy Cincinnati Milacron Co.)*

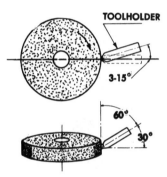

Figure 18-9 Diamond truing tools must be canted, as shown, to prevent chatter and gouging of the wheel to maintain sharpness of the diamond abrasive. *(Courtesy Cincinnati Milacron)*

- *Grain sizes*—Grain size in wheels range from 600 (fine) to 10 (coarse).
- *Type of bond*—Material that holds the wheel together.
- *Structure*—Structure of the grain in a wheel is designated by numbers from 1 to 12. The lower the number, the closer the grain spacing (Figure 18-10).
- *Grade*—Grades range from A to Z (soft to hard). Grades A through H are soft, grades I through P are medium, and grades Q through Z are hard.

The standard grinding wheel marking system is based on these factors (Figure 18-11).

There are many different grinding wheels used. Surface grinders are used in the grinding of precision parts (Figure 18-12 and Figure 18-13). They come in many different shapes to perform the wide variety of grinding operations required in industry.

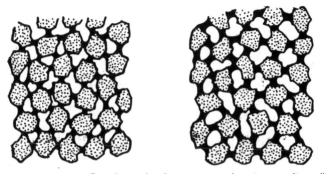

Figure 18-10   Grinding wheel structures showing medium (left) and wide (right) grain spacings. *(Courtesy Norton)*

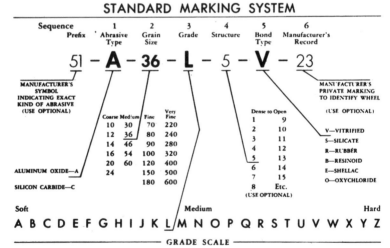

Figure 18-11   Standard grinding wheel system. *(Courtesy Aluminum Co. of America)*

Abrasive Blasting

Abrasive blasting is used when a quality surface finish is not of primary importance. Abrasive material is blasted on metal surfaces with great force. Usually, abrasive blasting takes place in a blasting room, on tables, or in blasting cabinets. The use of respirators is required for blasting in blasting rooms.

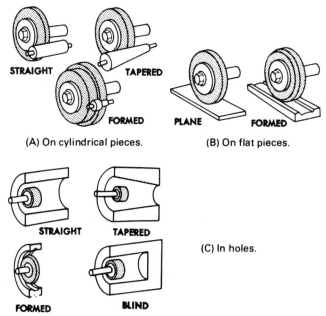

(A) On cylindrical pieces.

STRAIGHT TAPERED FORMED

(B) On flat pieces.

PLANE FORMED

STRAIGHT TAPERED

(C) In holes.

FORMED BLIND

Figure 18-12 Types of grinding commonly performed on external, internal, and flat work. Centerless grinding, either external or internal, is also performed on cylindrical work. *(Courtesy Cincinnati Milacron)*

Barrel and Vibratory Tanks

One of the most popular methods of finishing parts is called *barrel finishing*. Barrel finishing is also known as *abrasive tumbling*. Abrasive tumbling is a precision-controlled method of removing burrs, edges, and heat-treat scale from parts (Figure 18-14). This process also improves the surface finish of a part. A number of parts in the many different products that we use in our daily lives are finished with this process.

In the barrel-finishing process, abrasive media is put into a vibrating or rotating barrel along with water and a chemical compound. The action that takes place within the barrels removes the sharp burrs, edges, and heat-treat scale and then polishes the part (Figure 18-15).

Belt Sanders

Metal is also ground with belt sanders. Grinding requires the use of special safety equipment to protect the eyes and hands (Figure 18-16). It is a dangerous operation. Safety rules must be observed at all times.

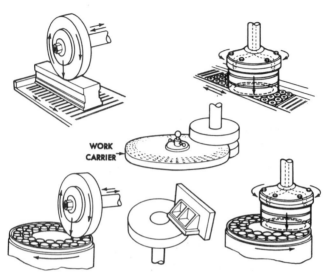

Figure 18-13 Principal types of surface grinding include flat and rotary tables and horizontal and vertical wheels. Disc sanders may be single- or double-wheel and either vertical or horizontal. *(Courtesy Cincinnati Milacron Co.)*

Figure 18-14 A typical part before (top) and after (below) vibratory barrel finishing.

(Courtesy Aluminum Company of America)

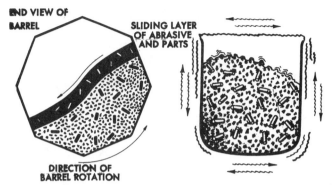

Figure 18-15   Barrel-finishing action within a rotating barrel (left) and vibratory barrel (right). *(Courtesy Norton)*

Figure 18-16   An operator of a belt sander finishing an aluminum casting. *(Courtesy Alcoa)*

Safety Rules for Abrasive Metal Finishing

Table 18-1 shows some safety tips when working with metal finishing machines.

Table 18-1   Safety Rules for Abrasive Metal Finishing

Category	Safety Check	Reason
Grinding wheels	Check for balance and ruts.	Unbalanced and rutted wheels can cause workpieces to be thrown.
	Operate at the proper speed.	Too high an operating speed can cause the wheel to burst.
	Check for worn center holes.	Worn center holes can cause the workpiece to be thrown.
Abrasive blasting	Check for secure hose connection.	Accidental disconnection of nozzle from hose can cause the hose to whip.
	Check for "dead man" switch.	No "dead man" switch to turn off blasting if hose is dropped could cause injury.
	Check for clear glass in blasting helmet window.	Poor visibility through blasting helmet window due to dirty or scratched glass could give serious problems to the operator.
	Check for excessive hose wear.	Excessive wear of hose can cause hose to break.
Barrel and vibratory tanks	Check for a lockout device.	Barrels and tanks can be accidentally turned on during loading unless they have a lockout device.
	Check for unguarded pinch points.	Barrels and tanks have many pinch points that may not be properly guarded.

Table 18-1 (continued)

Category	Safety Check	Reason
Belts	Check for cracked and torn belts.	Cracked or torn belts can whip around when broken.
	Check for unguarded belts.	Unguarded belts can allow you to come in contact with the belt.
	Operate at the proper speed.	Too high an operating speed can cause belts to break.

Summary

There are two major types of grinding: nonprecision and precision grinding. Nonprecision grinding is also called offhand grinding. Metal is removed by this method when there is no great need for accuracy. Pedestal and bench grinders are used for this type of work.

In precision grinding, metal can be removed with great accuracy. There are a number of different precision machines available that can grind metal parts to different shapes and sizes with very accurate dimensions.

The grinding wheels used in both nonprecision and precision grinding are made from aluminum oxide or silicon carbide. Abrasive blasting is used when quality surface finish is not of primary importance. Abrasive material is blasted onto metal surfaces with great force. The use of respiratory equipment is required for work done in blasting rooms.

Barrel and vibratory tanks are used for abrasive tumbling. Abrasive tumbling is a precision-controlled method of removing sharp edges, burrs, sand, and heat-treat scale from parts. The process also improves the finish of a part. Abrasive tumbling is also known as barrel finishing.

Metal is also ground with belt sanders. This type of grinding requires the use of special safety equipment to protect the eyes and hands. In fact, all the various types of abrasive metal finishing require that safety rules be observed at all times.

Review Questions

1. What are two types of grinding?
2. What is abrasive blasting?

3. What is the purpose of barrel finishing?

4. What types of work are finished on belt sanders?

5. What is another name for barrel finishing?

6. What are the nine standard shapes for grinding wheels?

7. Why should you never omit the blotting paper washers when mounting a grinding wheel?

8. Why must diamond truing tools be canted?

9. Disc sanders may be single- or double-wheel and either vertical or _____.

10. What is a "dead man" switch?

Chapter 19

Electroforming

The development of newer and tougher alloys and other materials has increased the demand for processes and machines that are able to machine these materials more quickly and economically. Many production items that were formerly produced on grinding and milling machines are now produced on electrochemical and electroforming machines.

Electrochemical Machining Process

Electrochemical machines are used for drilling, trepanning, and shaping extremely hard and tough materials (Figure 19-1). These machines may range from a small drill of the bench type to the larger automatic-cycle production machines.

Figure 19-1 Electrochemical machine. *(Courtesy Ex-Cell-O Corp.)*

365

Basic Principle

In the electrochemical machine, a low-voltage electric current is passed through a conductive fluid between the work and the tool electrode (see Figure 19-2). The fluid also flushes away the residue from the machining operations.

Electrochemical Machining (ECM)

Metal removal by electrochemical machining (ECM) is a process in which electrolytic action is used to dissolve the workpiece material. It is, in effect, the opposite of electroplating. The term *ECM* is sometimes used for the electrolytic grinding process, which is a modification of electrochemical machining.

Figure 19-2 shows the electrical diagram of an electrochemical machine. The workpiece (which must be a conductor of electricity) is placed in a tank on the machine table and connected to the positive terminal (anode) of a dc power supply. The tool electrode, which is shaped to form the required cavity in the workpiece, is mounted in the toolholder and connected to the negative terminal (cathode) of the supply. An electrolyte (liquid) flows through the gap between the tool and workpiece and is then pumped back to the working zone either through the tool or externally, depending on the application.

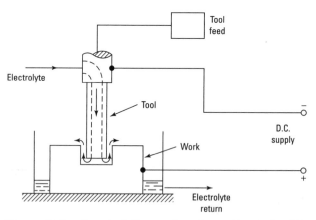

Figure 19-2   Electrical diagram for electrochemical milling.

Current flowing through the electrolyte dissolves the metal at the anode (+ terminal), that is, the workpiece. The electrical resistance is lower (and hence the current highest) in the region where the tool and work are closest together. Since the metal is dissolved from the work most rapidly in this region, the form of the tool will be reproduced on the work.

There is no mechanical contact between the work and the tool, and any tendency of the work metal to be plated on the tool (which is the cathode) is counteracted by the flow of the electrolyte that removes the dissolved metal from the working zone. Hence, there is neither tool wear nor plating of the work material on the tool, so one tool can produce a very large number of components in its life.

Machine Operation

Some machines may be equipped with push buttons and controls for manual cycling. The machines may also be equipped for either semi-automatic or automatic cycling to suit production requirements.

In the machine shown in Figure 19-3, a vertical ram supports a platen for mounting the tool electrodes. The platen is insulated from the ram and is connected through bus bars and cables to the electrical power supply. The power supply delivers current to the work area at low voltage.

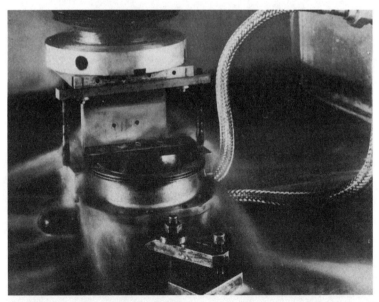

Figure 19-3   A conductive fluid carries the electric current that passes between the tool and the work.

The electrolyte system consists of a reservoir, a filter, a pump and drive motor, work compartment connections, and a return flow line. Corrosion-resistant materials (such as plastic or stainless steel) are used throughout the electrolyte system. The electrolyte is filtered before it enters the work area.

The type of electrolyte solution used depends on the nature of the materials to be cut and operations to be performed. Neutral, acidic, or alkaline solutions are used, depending on the operation.

Operations Performed

One advantage of the electrochemical process is that there is no tool wear. Any number of holes may be pierced without wearing away the electrode (Figure 19-4).

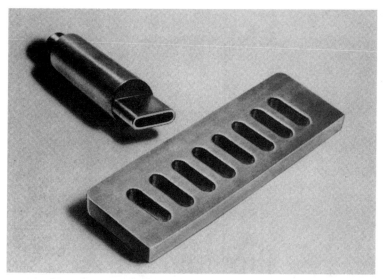

Figure 19-4 Elongated holes produced by the tool electrode.
(Courtesy Ex-Cell-O Corporation)

Internal gear shapes, as well as round holes and various other shapes, can be produced by the electrochemical process (Figure 19-5). Tool marks and burrs are completely avoided. Therefore, these do not have to be machined off, as in most other processes.

Figure 19-6 shows a die, and the electrode used to machine it. Three-dimensional cavities may be used in the process.

The electrochemical machine can be used as a general-purpose production machine. The process can be adapted to drilling, cutting

Figure 19-5 The production of internal gears by the electrochemical process. *(Courtesy Ex-Cell-O Corporation)*

Figure 19-6 A die (left) and the electrode (right) used to machine it. *(Courtesy Ex-Cell-O Corporation)*

off, shaping, trepanning, broaching, and cavity-sinking operations on any metal that is an electrical conductor.

Electrical Discharge Machining Process

This process can be used to machine hard or tough materials to any form that can be produced in an electrode. It can be used only on electrically conductive materials.

Basic Principle

Electrical discharge machining is performed with both the electrode and the work submerged in a dielectric fluid. The electrode and the workpiece must be in close proximity. Direct-current (dc) pulses at high frequency are discharged from the electrode through the resistance of the fluid. A minute particle is eroded from the surface of the work with each pulse.

Electrical Discharge Machining (EDM)

Electrical discharge machining (EDM) is a process that is sometimes referred to as electroerosion or electrospark machining. It is described as the removal of material by means of repetitive short-lived electric sparks that occur between the tool (called an *electrode*) and the workpiece.

The circuit used is a dc relaxation circuit that is fed from a mercury arc or selenium-type rectifier with more modern machines equipped with diodes. The latest equipment uses less expensive semiconductors to produce the dc power needed for the process.

To achieve erosion, a gap of approximately 0.025 to 0.05 millimeters between the tool and workpiece has to be maintained. This is done by the servo drive. This type of circuit is not too efficient but does represent the earlier type of circuit used for EDM.

In Figure 19-7 note how the capacitor is placed across the tool and workpiece. When the power is turned on, the voltage begins to build up and charge the capacitor (C). In this initial stage, the spark gap behaves as an open circuit and no current flows. As the voltage across the capacitor builds up, it eventually reaches the breakdown voltage of the gap. The breakdown voltage required for the gap is determined by the distance between the electrode and the workpiece, as well as the dielectric fluid used. A spark is produced across the gap when the voltage reaches the critical point. The dielectric ionizes, and the capacitor is discharged. After the capacitor has discharged, the surrounding dielectric deionizes and again becomes an effective insulator. The cycle is then repeated. Thus, a rapid succession of sparks is obtained. The frequency of spark production is on the order of 10,000 per second.

In high-production applications, the high-intensity, low-frequency spark production of the relaxation circuit is unsatisfactory. The sparks of a higher frequency are used for high production. The spark is produced by a rotary impulse generator. This ensures better control of frequency and higher metal removal rates, and, in addition, less tool wear occurs.

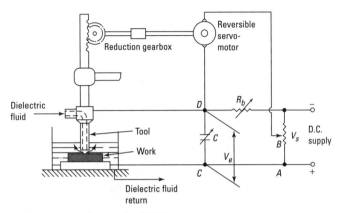

Figure 19-7   Circuit for EDM.

Machine Operation

The electroforming machine shown in Figure 19-8 is equipped with an electrohydraulic servo for maintaining a constant gap between the electrode and the work. The servo automatically feeds the tool at the proper rate to keep the gap constant as the cutting progresses.

The power supply feeds the current to the servo until the electrode reaches the preset depth, and a predetermined amount of stock is removed. The power supply controls the movement of the electrohydraulic servo unit in the machine tool while precisely metering high-frequency dc pulses to the electrode. The machine tool is a precision mechanical device. It holds the work and feeds the electrode to the work while maintaining an extremely precise gap. The servo automatically retracts the tool if the gap becomes too small or if residual from the cutting operation forms a bridge. Momentary retraction of the tool clears the cutting area, and the servo resumes its downfeed. Accurate gauging and depth of machining are controlled by the combination of a micrometer and limit switches.

An extremely uniform gap is held between the tool and work, regardless of configuration. Because of the extreme repeatability and uniformity of the process, extremely accurate predictions of results are possible. Size, tolerances, and surface finish of the work can be predicted within very close limits.

Cutting speed is controlled by amperage, or cutting current. Surface finish of the work is controlled by the frequency at which the cutting current pulses occur.

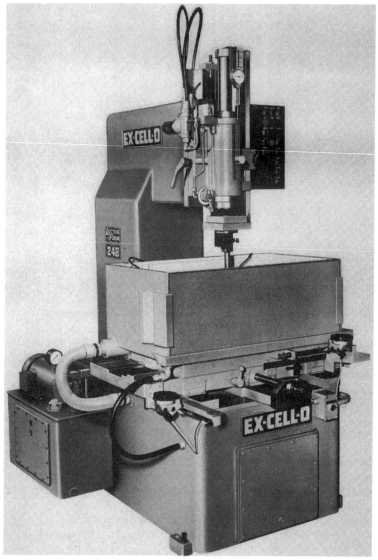

Figure 19-8 This machine uses the electrical discharge machining (EDM) process of forming metal. *(Courtesy Ex-Cell-O Corporation)*

A workpan is provided on the machine for setup of the work. An adjustable level control and an adjustable safety float switch are provided to control the level of the dielectric fluid.

Operations Performed

Electrical discharge machining is particularly adaptable to general-purpose machining and for die work. Various types of work may be done on these machines.

Figure 19-9 shows a multiple-electrode setup, where 18 holes are drilled in an aircraft fuel nozzle in a single machining cycle.

Figure 19-9   A multiple-electrode setup (left) in which 18 holes are drilled in the aircraft fuel nozzle part (right) in one machining cycle.
(Courtesy Ex-Cell-O Corporation)

Figure 19-10 illustrates some of the miscellaneous forms that can be machined by the process. Hardened steels and carbides as well as tough alloys can be machined without difficulty (Figure 19-11).

Electroforming can be used to greatest advantage in die manu-facturing and machining operations involving intricate forms that are difficult to produce by conventional machines. This is particu-larly advantageous in machining of parts in the hardened condi-tion (Figure 19-12). Dies can be reworked, and reworking can be performed earlier to maintain a high quality-control level in the production of parts.

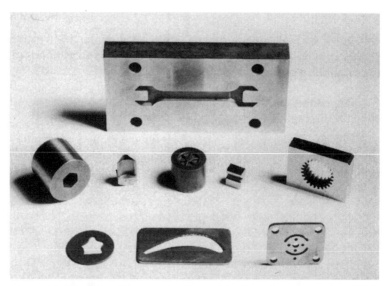

Figure 19-10 Miscellaneous forms machined by the electrical discharge machining process. *(Courtesy Ex-Cell-O Corporation)*

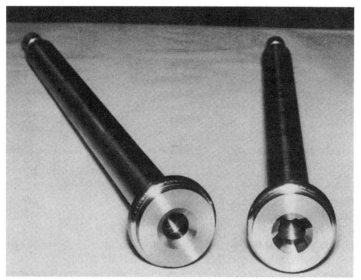

Figure 19-11 A single-axis movement of the electrode machined this four-lobed spline in the stainless steel shaft. *(Courtesy Ex-Cell-O Corporation)*

Figure 19-12 Component parts of a die that is used to produce powdered metal cluster gears, showing internal involute gear form rough and finish machined in the hardened condition. *(Courtesy Ex-Cell-O Corporation)*

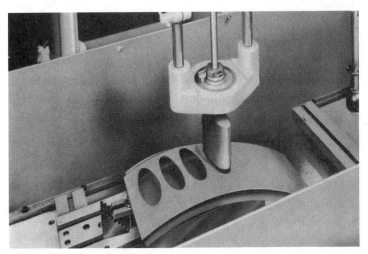

Figure 19-13 A typical example of an irregular shape easily cut in tough material by electrical discharge machining. This is a stainless steel shroud, pierced by a brass electrode positioned above it. *(Courtesy Ex-Cell-O Corporation)*

Irregular shapes can be cut easily in tough materials (Figure 19-13). Forms can be produced with a soft tool. Tool pressure (which can cause localized stress) and heat (which can cause distortion) are both absent in electrical discharge machining.

Figure 19-14 illustrates other items that may be produced by the electroforming process more quickly than by conventional means. Tools used on conventional machines may be fitted on the electroforming machine, especially after they have been heat-treated.

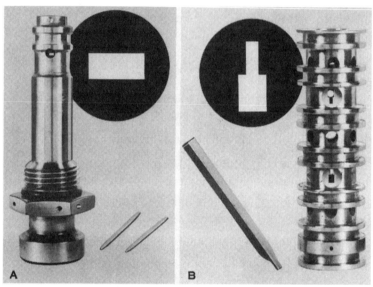

Figure 19-14   Valves produced by the electroforming process: (A) A high-pressure relief valve (insert shows the shape of the opening); (B) another valve; (C) shape of the opening is shown in the insert.
(Courtesy Ex-Cell-O Corporation)

Summary

Electrochemical machines are used for drilling, trepanning, and shaping extremely hard and tough materials. In the electrochemical machine, a low-voltage electrical current is passed though a conductive fluid between the work and the tool electrode.

One big advantage of electrochemical machining (ECM) is that there is no tool wear. Many holes can be pierced without wearing away the electrode. Tool marks and burrs are completely avoided.

Therefore, these do not have to be machined off, as in most other processes.

Electrical discharge machining (EDM) is performed with both the electrode and the work submerged in a dielectric fluid. Direct-current pulses at high frequency are discharged from the electrode through the resistance of the fluid. Minute particles are eroded from the surface of the work by each pulse. There is no mechanical contact between the work and the tool, and any tendency of the work metal to be plated on the tool (which is the cathode) is counteracted by the flow of electrolyte that removes the dissolved metal from the working zone. This means there is no tool wear or plating of the work material on the tool.

EDM is sometimes referred to as electroerosion or electrospark machining. It is sometimes described as the removal of material by means of repetitive short-lived electric sparks that occur between the tool (called an electrode) and the workpiece. The circuit used is a dc relaxation circuit that is fed from a mercury arc or selenium-type rectifier. In order to work properly, a gap of approximately 0.025 to 0.050 mm between the tool and workpiece must be maintained.

Review Questions

1. How does industry employ electrochemical machines?
2. What are the advantages of electrochemical processing?
3. What is the principle of operation of the electrical discharge machine?
4. What is electroforming?
5. What is electrochemical machining?
6. What does the term *cathode* mean?
7. What does the term *anode* mean?
8. What is an electrolyte?
9. What type of circuit is used for EDM?
10. How is cutting speed controlled in EDM?

Chapter 20

Ultrasonics in Metalworking

Ultrasonics, using a form of cavitation, may be utilized in several different machining processes. Drilling, grinding, cutting, and soldering or welding may all be performed with ultrasonics.

Machining

The machine processes use an abrasive slurry, which is pounded against the material at an ultrasonic rate, causing it to be chipped away. Since the work is not chipped, heated, stressed, or distorted, a greater degree of precision may be obtained than with conventional machining methods. The size of the abrasive grit used determines the degree of fineness of cut; coarse abrasive grit cuts the material at a faster rate. Ultrasonic machining is especially advantageous in cutting very hard materials such as glass, ceramics, steel, and tungsten. It is practically ineffective on soft substances.

Abrasives

The abrasive used in ultrasonic machining is suspended in a liquid. This liquid serves several functions. It provides an acoustic bond between the vibrating tool and the work. Therefore, it produces an efficient way to transfer energy from the tool to the work. It can also act as a coolant at the tool face. Another task it performs is to carry the abrasive to the cutting area and then wash away the spent abrasive and swarf. *Swarf* is the filings and the wet materials left over after the abrasive has done its work.

Abrasives used for ultrasonic machining are many, but those most commonly used are aluminum oxide, silicon carbide, boron carbide, and some diamond dust. Grit sizes vary between 100 and 2000. The coarse grade is used for roughing, and the finer grades (600 to 700) are used for finishing. Very fine grades (1000 to 2000) are used for the final pass on highly accurate work.

Liquid used to carry the abrasive must have the following properties:

- Its density must be approximately equal to that of the abrasive.
- It must have good wetting properties.
- To be able to carry the abrasive down between the tool and the work, it must have low viscosity.

- Efficient removal of heat from the cutting zone requires the liquid to have a high thermal conductivity and specific heat. Water can be used for the liquid if it has rust inhibitors and wetting agents added.

Drilling

An ultrasonic drill concentrates all the vibrations at a specific point, thereby making use of all the available energy at a single point. In concentrating the energy, the amplitude of the vibrations is increased. This is brought about by the use of a tapered stub (Figure 20-1). In the cross-sectional view, note that the driving coils are wound around the magnetostrictive nickel core. The laminations are firmly connected to the tapered stud, which extends out of the drill housing. As the core changes size, it causes the stub to vibrate up and down to set up longitudinal vibrations, thereby creating the drill motion. The assembly has an inlet and outlet for circulation of water to provide cooling, since the vibrations produce heat.

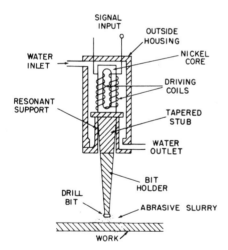

Figure 20-1 Cross-sectional view of an ultrasonic drill.

The drill tip actually moves up and down only a very short distance, but the rapid rate of vibration makes it very effective in driving through hard materials. The vibrating effect combines with cavitation in the liquid-base abrasive to perform the driving action. The tip of the drill is applied to the abrasive, which performs the actual drilling action. Aluminum oxide, boron carbide, silicon carbide, Carborundum, diamond dust, or similar hard materials mixed with oil or water can be used as the abrasive. A container for the abrasive is mounted so that it can be applied at the point where the drilling is being performed.

Most ultrasonic drilling machines resemble rotary drill presses in appearance. A commercial ultrasonic drill (Figure 20-2) is arranged so that different bits can be substituted in the same transducer assembly. Since drilling is by reciprocating action (rather

Figure 20-2 An ultrasonic drill. At the lower left, notice the cuts already made. *(Courtesy Gulton Industries)*

than by rotary motion), each cutting tool is made in the same shape as the hole to be drilled. A rotary drill can drill only round holes, but an ultrasonic drill can cut holes in almost any desired shape. With a properly constructed tool, a large number of holes can be machined in a single operation. Drill bits are made of cold-rolled or unhardened steel and then silver-soldered to the core-shaped shaft.

Grinding and Cutting
Grinding and cutting operations by ultrasonics on hard materials are similar to drilling. Grinding involves either shaping a material to a particular form or smoothing it to a particular finish. A coarse abrasive is used at the start; then, a finer grain is used until the correct finish is obtained. Ultrasonic grinding has several distinct advantages, as compared with conventional grinding. A good finish may be obtained at a lower temperature, thereby reducing damage. The amount of grinding can be controlled accurately and a good finish is possible on coarse materials. Figure 20-3 shows an ultrasonic grinder.

Ultrasonic grinding and cutting are also advantageous in that they work best with harder materials, while ordinary machine tools work best with the softer materials. Ultrasonics can be used to cut glass, quartz, sapphire, germanium, silicon, diamond, ceramics, and other hard nonmetallic substances. The various hard materials can be cut or sliced into thin wafers for various uses. With a properly shaped tool, a number of slices can be cut at one time (Figure 20-4). Cutting blades are usually made of thin steel or molybdenum and can be used to cut wafers as thin as 0.015 inch.

Cutting Performance
Just how the process works has not yet been fully understood or explained. People working in the field do not agree as to how the cutting action takes place. One claim is that the cutting rate increases linearly with an increase in three variables and that, above a certain critical abrasive/water ratio, the relationship R/afd is constant, as defined here:

R = penetration rate of tool
a = amplitude of vibration
f = frequency of vibration
d = mean diameter of abrasive

Figure 20-3 The head unit of an ultrasonic grinder. *(Courtesy Raytheon)*

However, others say there is a nonlinear relationship between amplitude, frequency, and cutting rate. Factors such as an excessive force on the tool may prevent abrasives reaching the tool so that cutting would eventually stop and cause a nonlinear condition to exist. The position of the maximum cutting rate was influenced by the amplitude of the tool and the cross-sectional area of the tool. Shape of the tool face could influence the maximum cutting rate. A narrow rectangular cross section gives a higher maximum cutting rate than a tool having a square cross section of the same area.

The viscosity of the slurry has a damping influence on the cutting tool oscillations. This is of importance when manufacturing

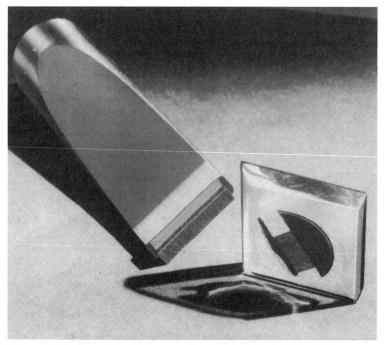

Figure 20-4   An ultrasonic cutter head. A large number of slices can be produced with each cut.

work with a large side area (such as with splines). If too low, the viscosity will retard passage of slurry and waste the side clearance. This, too, reduces the cutting rate and can be eliminated if the shank of the tool is reduced in size behind the face.

Abrasive cutting success depends not only on its hardness but also on the number and durability of the cutting edges that are a function of the shape of the particles. The brittle behavior of the work material greatly influences the cutting rate. Brittle, nonmetallic materials can be cut at very high rates.

The process has one important advantage. It leaves no residual mechanical stresses in the cut material when operating under normal conditions. It can be shown that, even with the most easily fractured materials, the chipped workpieces are small even when compared with the abrasive, indicating that the stresses transferred to the work must be very localized. It is important to maintain an adequate supply of liquid coolant.

Advantages of Ultrasonic Machining

It is easy to obtain accuracy and good surface finish. There is no heat generated, which means there are no changes in the physical structure of the material.

The equipment is safe to handle and requires little skill to operate. Cheap abrasives are used. The process can be used with all brittle materials and is not limited by any other physical property.

Newer Developments

Newer techniques make use of a diamond-impregnated cutting tool. The tool has a rotary transducer that gives much faster cutting rates and greater accuracy than the conventional methods. As the cutter rotates, the process is said to be similar to that of end milling. So, by traversing the tool, profiles can be generated. Work tolerances in the region of 0.0125 millimeters in glass and ceramic components are possible. When machining glass, the cutting rate is on the order of 10 cubic millimeters per minute. The tool is rotated at 100 to 1800 revolutions per minute with an axial vibration in the region of 20 Hz and an amplitude of 0.005 to 0.0125 millimeters.

Soldering and Welding

When metals such as copper or aluminum are exposed to the atmosphere, a thin film of oxide forms on the metal. Thus, the film prevents the solder from making proper contact with the metal. A chemical flux can be used to coat copper materials to prevent formation of the oxide, but this will not work well on aluminum. Application of ultrasonic vibrations to molten solder, when in contact with aluminum, cavitates the solder and eliminates the oxide film. Thus, the solder can reach and contact the surface of the aluminum. Copper and other metals can be soldered in the same manner, using solder with a high tin content. After tinning by ultrasonic means, soldering can be done by the usual methods. Ultrasonic soldering usually requires no flux, even for joining dissimilar metals. Figure 20-5 shows one type of ultrasonic soldering iron.

Ultrasonic welding differs considerably from other types. No external heat is applied—only ultrasonic vibrations—but similar and dissimilar metals can be joined effectively. Aluminum can be welded either to aluminum or to another metal. Stainless steel, molybdenum, titanium, and other metals can be joined either by spot welding or by seam welding.

The amount of heat created by the ultrasonic vibrations is not sufficient to melt the metals being joined. Therefore, the bond

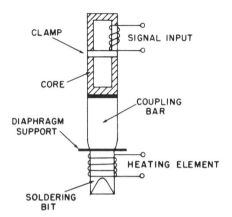

Figure 20-5 An ultrasonic soldering iron.

occurs by some other means (the joining process is not completely understood). Evidently, the two sections are joined by a molecular bond formed by the combined action of the heat and the fast rate of vibration. As in other types of welds, the bond is stronger than either of the metals involved. Special preparation of materials is not required in ultrasonic welding, and much less pressure is required

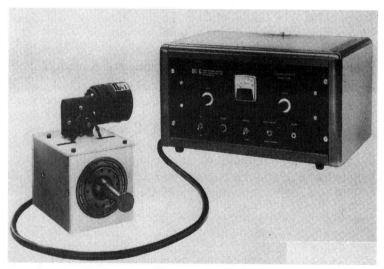

Figure 20-6 An ultrasonic seam welder. The motor turns the transducer head. *(Courtesy Gulton Industries)*

to hold the sections together during the process. Since there is so little heat, there is less deformation of the materials being welded.

One type of ultrasonic welding unit (Figure 20-6) resembles an ultrasonic drill. It has a tapered stub with a large tip on the end. The vibrations are applied parallel to the surfaces of the materials being welded. These are shear waves; longitudinal waves will not weld. A seam weld is formed by running the tool down strips of metals. A spot weld is produced by touching the tool at a definite point.

Summary

Ultrasonics may be utilized in several different machining processes. Drilling, grinding, cutting, and soldering or welding may be performed with ultrasonics. The machine process uses an abrasive slurry, which is pounded against the material at an ultrasonic rate, causing it to be chipped away.

An ultrasonic drill concentrates all the vibrations at a specific point, thereby making use of all the available energy at any single point. The drill tip actually moves up and down only a very short distance, but the rapid rate of vibration makes it very effective in driving through hard materials. The drill tip actually moves so rapidly that it effectively drives through hard materials. This vibrating effect combines with cavitation in the liquid-base abrasive to perform the driving action.

Grinding and cutting operations by ultrasonics on hard materials are similar to drilling. A coarse abrasive can be used at the start; a finer grain is used until the correct finish is obtained. A good finish can be obtained at a lower temperature, thereby reducing damage. Ultrasonics can be used to cut glass, quartz, sapphire, germanium, silicon, diamonds, ceramics, and other hard materials.

Ultrasonic welding and soldering differ considerably from the regular process. No external heat is applied, only ultrasonic vibrations. Similar and dissimilar metals can be joined effectively. The amount of heat created by the ultrasonic vibrations is not sufficient to melt the metals being joined. Therefore, the bond occurs by other means.

Review Questions

1. How does the ultrasonic process work?
2. What are the advantages of ultrasonic welding?
3. How is cutting and grinding accomplished in the ultrasonic process?

4. What abrasives are commonly used in ultrasonic machining?

5. What are the needed properties of the liquid used to carry the abrasive away from the work area in ultrasonic milling?

6. What do most ultrasonic drilling machines resemble?

7. The grinding and cutting operations performed by the ultrasonic process on hard materials are similar to _____.

8. What effect does the viscosity of the slurry have on the cutting tool?

9. What is the main advantage of ultrasonic machining?

10. A _____ weld is produced by touching the tool at a definite point.

Chapter 21

Machine Shop Robotics and Electronics

The machine shop has been modernized to the extent that robotics and electronic controls have been introduced into the operation and control of machines. In most instances, this has increased efficiency, improved the quality of the product, and improved production numbers. The demand for high-precision products has also been answered in work for the aerospace industry and certain control devices such as linear-motion controls and tables.

Applications

Robots can be used in the machine shop to drill, route finished products, deburr, hone, lap, grind, and do other procedures. For example, Cincinnati Milacron Company's machining system can do accurate positioning for demanding process applications. The computer numerical control (CNC) drill provides automatic process control and drilling precision for various machining operations. The pneumatic router can cut a wide variety of materials. The long string brush deburring system can remove burrs from many different materials without requiring accurate fixturing. A flexible interface enables the robot to monitor the status of the system equipment. Figure 21-1 shows the location of the various parts of the *Atlas Robot* used in the machine shop.

Robots

A robot is made up of a number of subsystems. They can be classified in a number of ways, but the only one discussed here is the *industrial robot.*

The industrial robot has arms and grippers attached (Figure 21-2). The grippers are fingerlike and can grip or pick up various objects. They are used to pick and place, as well as move materials precisely where they are needed for machining, grinding, or finishing. The robot can be programmed and computerized. The teach box is used to program the microprocessor used as the computer brain (Figure 21-3). Sensory robots, welding robots, and assembly robots usually have a self-contained computer.

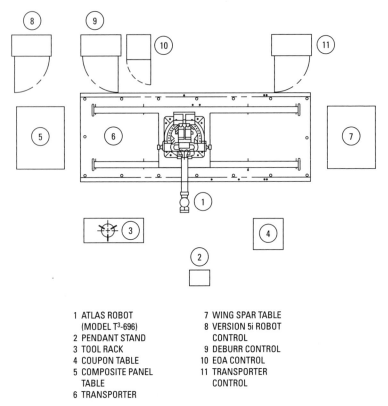

1 ATLAS ROBOT
(MODEL T³-696)
2 PENDANT STAND
3 TOOL RACK
4 COUPON TABLE
5 COMPOSITE PANEL
TABLE
6 TRANSPORTER

7 WING SPAR TABLE
8 VERSION 5i ROBOT
CONTROL
9 DEBURR CONTROL
10 EOA CONTROL
11 TRANSPORTER
CONTROL

Figure 21-1 Robotic machining system. *(Courtesy ABB)*

Basic Construction

The basic components of the robot include the *manipulator*, the *base*, the *arm*, the *wrist*, and the *grippers*. Figure 21-4 shows a complete robot system.

Manipulator

The manipulator is one of the basic parts of the robot (Figure 21-5). All components of a robot must be operational for the robot to function and do its job accurately. The manipulator is classified by certain arm movements. There are, for instance, four coordinate systems used to describe the arm movement: polar coordinates, cylindrical coordinates, Cartesian coordinates, and articulated (jointed-arm, spherical) coordinates.

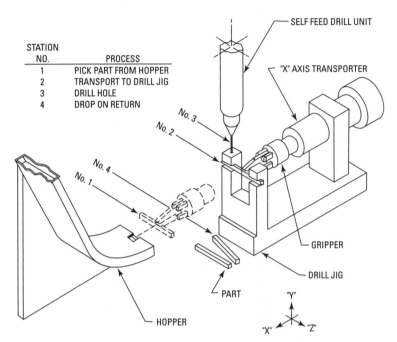

STATION

NO.	PROCESS
1	PICK PART FROM HOPPER
2	TRANSPORT TO DRILL JIG
3	DRILL HOLE
4	DROP ON RETURN

SELF FEED DRILL UNIT

"X" AXIS TRANSPORTER

No. 3

No. 2

No. 4

No. 1

GRIPPER

DRILL JIG

PART

"Y"

"X" "Z"

HOPPER

Figure 21-2   Robot with gripper.

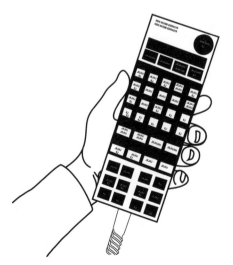

Figure 21-3   A teach
pendant. *(Courtesy ABB)*

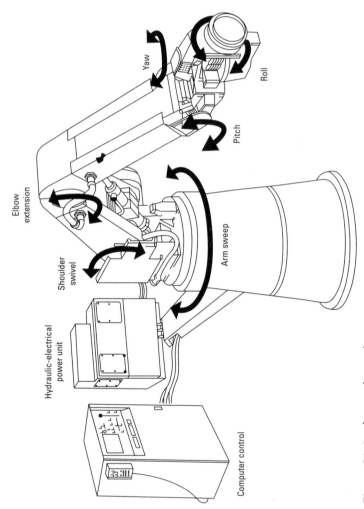

Elbow
extension

Yaw

Roll

Pitch

Shoulder
swivel

Arm sweep

Hydraulic-electrical
power unit

Computer control

Figure 21-4 A complete robot system. *(Courtesy ABB)*

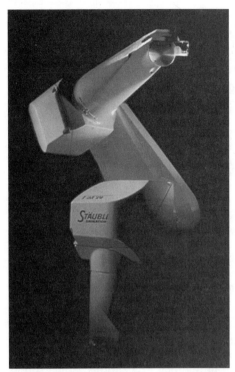

Figure 21-5   Manipulator in a robot system.
(Courtesy STÄUBLI CORPORATION)

Base

The robot is anchored to the floor or table at the base. The base may be rigid, or it can be designed as a supporting unit for all the component parts of the robot. The base does not have to be stationary, since it may become part of the robot. It may be capable of any combination of motions, including rotation, extension, twisting, and linear. Most robots have the base anchored to the floor, although, because of limited floor space, they may be anchored to the ceiling or to suspended support systems overhead. A track or conveyor system may be used to move the robot along as needed.

Arm

Most industrial-type robots have some type of arm. It may be jointed and resemble a human arm, or it may be a slide-in/slide-out type used to grasp something and bring it back closer to the robot. A jointed arm consists of a base rotation axis, a shoulder rotation axis, and an elbow rotation axis. This type of arm provides the largest working envelope per area of floor space of any design thus far. If this is a six-axes type of arm, it requires some rather sophisticated computer control. Most arms now have some type of joint. From one to six jointed arms may be attached to a single base for special jobs. The expense of controlling this type of movement is rather large because of its complexity.

Wrist

The wrist is attached to the jointed arm. The wrist is similar to a human wrist and can be designed with a wide range of motion, including extension, rotation, and twisting. This aids the robot in reaching places that are hard to reach by the human arm (such as inserting a machined piece into a furnace or in heat treating). It comes in handy especially when spray-painting the interior of an automobile, working on the assembly line, and welding inside a pipe. This type of flexibility will improve the machining of products, as well as assist in loading and unloading a milling machine or a drill press.

Grippers

The grippers are at the end of the wrist. They are used to hold whatever the robot is to manipulate. Pick-and-place robots have grippers to move objects from one place to another. Some robots have end-of-arm tooling instead of grippers. In such cases, the robot is used primarily for one type of operation (such as spray-painting or welding). If a tool is attached, it is unnecessary to have a gripper on the end of the arm. A pneumatic impact wrench can be fitted at the end of the arm just as easily as grippers. Figure 21-6 shows a piezoresistive transducer on the end effector being used to monitor pneumatic pressure for cost-effective control of an automated drilling robot. Grippers are made in a number of sizes and shapes. Various types are made to fit the job being done by the robot such as that shown in Figure 21-6.

Motion

To move the manipulator, the robot needs some type of power source. There are three types of robot power sources: electric, pneumatic, and hydraulic. Each has its advantages and disadvantages. For example, the electric motor drives cannot lift as much as hydraulic drives. However, electricity can be more precisely controlled, so this type is usually relied upon for very fine work.

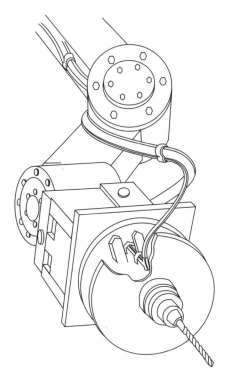

Figure 21-6 Automated drilling with a drill chuck end effector.

Workshop Duties

Robots can be used as part of a work cell. They perform simple, repetitive materials-handling tasks with ease. Transferring, loading, and unloading are simple applications, but they are tasks that can free humans for more productive work. Robots can be efficient links between different machines, as well as between machines and material supply sources (Figure 21-7). Robots provide efficiency that pays for itself with increased production.

Figure 21-8 shows how one machine is used to feed another and thereby relieve a human operator from the drudgery of repetitive work.

Robots can be used effectively to do spot welding, spray-painting, machine tool loading, materials transfer, investment casting, forging, palletizing, glass handling, and press loading, as well as stacking and destacking.

The machine shop worker must be able to make the various grippers and other parts of the robot. Precision workmanship

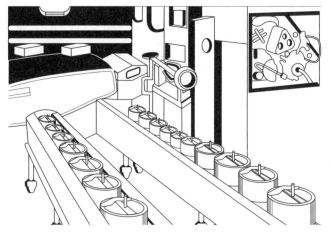

Figure 21-7 Robot tending two multispindle machine tools in the manufacture of cylinder heads. *(Courtesy Prab Robots)*

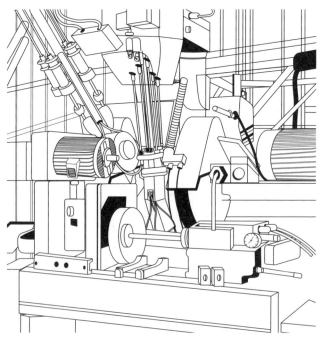

Figure 21-8 Machine feeding machine.

makes for fewer problems later in the routine operation of the robot. Some of the repetitive jobs in a machine shop can be handled very well by the robot or robotlike devices.

Computerized Measurements with Ultra-Digit

As the demand for greater accuracy and the ability to improve quality of the finished product increases, the need for more efficient and reliable means for checking the output of the machine shop becomes apparent. Manufacturers make specialized instrumentation for various quality-control and production requirements. The Ultra-Digit computer system serves as one example of the many approaches to improving the quality of today's machine shop production.

The Ultra-Digit computer system shown in Figure 21-9 has a statistical analysis (Stat-An) software program built into the system. This program and equipment are shown as an example of what is available. There are many other types with more advanced features available. The metals field (machine shops in particular) will be constantly updated with new and interesting ways to electronically improve the operation and testing of manufactured products produced by the machine shop and its machines. The operator of this system has no need to understand computer programming. However, additional programs can be created using the well-known BASIC language.

The modular system shown in Figure 21-9 permits simple printouts of each measurement displayed by simply using the master interface, a connecting cable, and printer. Up to eight Ultra-Digits can be connected to the master interface.

There are three additional auxiliary interfaces to the master interface. Each auxiliary permits hooking up an additional eight Ultra-Digit units. Printouts of the measurements of 32 or 64 Ultra-Digit indicators are possible with this method.

Add the Ultra-Digit computer to run the built-in Stat-An program. Stat-An for the standard system permits use of up to 32 indicators and can be used with a TV or video monitor. A later version of the statistical program is also available. It allows use of up to 64 indicators without using a TV or video monitor.

System Components

Components for this system are identified in Figure 21-10. The following numbers indicate the component:

1. Ultra-Digit indicator
2. Lifting lever
3. Gauge stand

Figure 21-9 The Ultra-Digit computer system. *(Courtesy Fred V. Fowler)*

   4. Foot switch
   5. Printer
   6. Computer
   7. Master interface
   8. Auxiliary interfaces
   9. Monitor or television

System Printouts
The software package is a menu-driven program that prompts the
user with directions and does not necessarily require a highly

Figure 21-10   Identification of component parts of the Ultra-Digit
computer system. *(Courtesy Fred V. Fowler)*

trained computer technician to operate. The operator has no need
to understand programming.

The software package is contained on a read-only memory
(ROM) chip in the master interface and is activated through the
computer system. It will monitor from 1 to 32 indicators and output directly to the printer, the television/monitor, or both.
Additional ROM chips can be added in the future for various programs, but the computer is programmable in BASIC for current
programming needs.

Figure 21-11 shows the sample output from the printer. Each part has been numbered and can be easily identified.

- No. 1—The automatic printout headings.
- No. 2—Select Channel allows user to select measuring mode (actual or nominal); select inch or metric; enter component tolerance or maximum/minimum size; enter predetermined lot size.
- No. 3—This shows that an entry of sample measurements came directly from the unit's indicators. Up to 99 measurements can be taken on a single channel, depending on the total number of channels. (*Symbols*: H = above tolerance; L = below tolerance.)
- No. 4—Note the number of fliers (any reading that is 100 percent of the component tolerance away from the sample average), the number of readings that are out of tolerance, percentage of readings above and below tolerance, percentage of tolerance used, max/min readings, range, average (mean), standard deviation, and ± 3 and 4 standard deviation values.
- No. 5—This histogram of all readings for this lot sample indicates minimum and maximum tolerance.
- No. 6—The X-Bar Chart: (average value) shows between upper and lower *tolerance* limits, and upper and lower *control* limits. This is a continuous process for the channel and component selected while the computer is switched on. (*Symbols*: X = normal reading; F = flier in calculations; E = upper control limit less than lower control limit.)
- No. 7—The range value of each sample is plotted automatically in this R Chart and is shown against the upper control limit. This is a continuous process for the channel and component selected while the computer is on.

Note
The computer remembers previous X-Bar and R readings and plots them together with the latest readings from the last sample.

Figure 21-12 shows the complete 32 Ultra-Digit system with all components except the foot switch. The system may be configured without the computer, thereby still having 32 indicators on-line, but the program cannot be run. Only real-time measurements can be captured and printed. Following are the components shown in Figure 21-12:

- 1–3—Auxiliary interfaces
- 4—Master interface with eight Ultra-Digit indicators

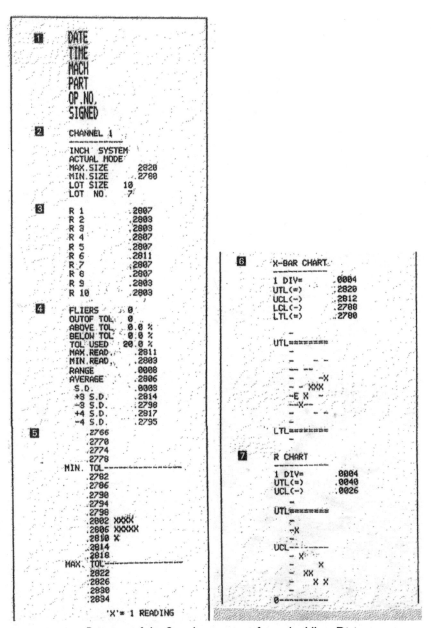

Figure 21-11 Printout of the Stat-An program from the Ultra-Digit computer system. *(Courtesy Fred V. Fowler)*

- 5—Computer
- 6—Television or video monitor
- 7—Printer

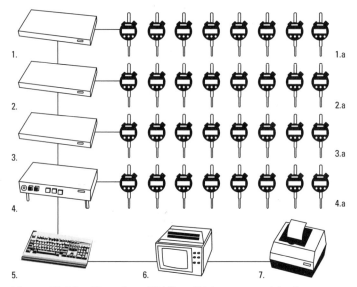

Figure 21-12 Complete 32 Ultra-Digit system with all components except the foot switch.

The standard system can drive a maximum of 32 indicators and will allow the use of a television or video monitor for faster results. For applications where more than 32 indicators are desired, a special version of the software program is needed. The second program will drive up to 64 indicators simultaneously. This later version, however, does not permit the use of a television or video monitor. Even later modifications will handle all types of outputs and peripherals.

Summary

Most machine shops have been modernized to the extent that robotics and electronic controls have been introduced into the operation and control of the machines. In most instances, this has increased efficiency and quality of the product and has improved production numbers.

Robots can be used in the machine shop to drill, route finished products, deburr, hone, lap, grind, and do other procedures. Robots are made up of a number of subsystems. They can be classified in a number of ways. Industrial robots have arms and grippers attached. They are made up of a manipulator, a base, and an arm with a wrist.

There are three types of robot power used to do the work demanded of it. There are the electrically powered, the hydraulically powered, and the pneumatically powered types of robots. Electric motors can lift limited amounts of load, while the hydraulically powered motors can lift heavy loads. Robots can be used as part of a work cell where they handle simple, repetitive materials-handling tasks with ease. Robots can also be used to do spot welding, spray-painting, machine tool loading, materials transfer, investment casting, forging, palletizing, glass handling, and press loading, as well as stacking and destacking.

The computer is used for more than directing the robot. Computer programs aid the machinist in producing accurate work, as well as improving output. Computer programs are available that allow the machine shop operator to easily transfer knowledge to the new system with little or no relearning required. The software package is contained within the computer and is menu-driven. It prompts the user with directions and does not necessarily require a highly trained computer technician to operate. The operator has no need to understand or do programming.

Review Questions

1. What is the robot's role in the machine shop?
2. What does CNC mean?
3. What is a robot?
4. Identify the following:
   A. Manipulator
   B. Base
   C. Arm
5. What do the grippers do on a robot?
6. How does the wrist work and for what purpose?
7. What is a work cell?
8. What kinds of operations can a robot perform?
9. What does *pick and place* mean?
10. What is a teach pendant?

Appendix

Reference Materials

This appendix contains useful reference information, including the following:

- Colors and approximate temperatures for carbon steel
- Nominal dimensions of hex bolts and hex cap screws
- Nominal dimensions of heavy hex bolts and heavy hex cap screws
- Nominal dimensions of heavy hex structural bolts
- Nominal dimensions of hex nuts, hex thick nuts, and hex jam nuts
- Nominal dimensions of square-head bolts
- Nominal dimensions of heavy hex nuts and heavy hex jam nuts
- Nominal dimensions of square nuts and heavy square nuts
- Nominal dimensions of lag screws
- American Standard machine screws (heads may be slotted or recessed)
- American Standard hexagon socket, slotted headless, and square-head setscrews
- American Standard cap screws (socket and slotted heads)
- English to metric conversion table
- Decimal equivalents, squares, cubes, square and cube roots, and circumferences and areas of circles (from $\frac{1}{64}$ to $\frac{5}{8}$ inch)
- Decimal equivalents, squares, cubes, square and cube roots, and circumferences and areas of circles (from $\frac{41}{64}$ to 1 inch)
- Number and letter sizes of drills with decimal equivalents

Colors and Approximate Temperatures for Carbon Steel

Color	Temperature in Degrees Fahrenheit	Temperature in Degrees Celsius
Black red	990	532
Dark blood red	1050	566
Dark cherry red	1175	635
Medium cherry red	1250	677
Full cherry red	1375	746
Light cherry, scaling	1550	843
Salmon, free scaling	1650	899
Light salmon	1725	946
Yellow	1825	996
Light yellow	1975	1080
White	2220	1216

Nominal Dimensions of Hex Bolts and Hex Cap Screws

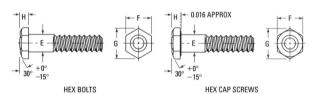

HEX BOLTS

HEX CAP SCREWS

Hex Bolts		Hex Cap Screws	
Nominal Size E	Width Across Flats F	Width Across Corners G	Head Height H
$1/4$	$7/16$	$1/2$	$11/64$
$5/16$	$1/2$	$9/16$	$7/32$
$3/8$	$9/16$	$21/32$	$1/4$
$7/16$	$5/8$	$47/64$	$19/64$
$1/2$	$3/4$	$55/64$	$11/32$
$5/8$	$15/16$	$1\,3/32$	$27/64$
$3/4$	$1\,1/8$	$1\,19/64$	$1/2$
$7/8$	$1\,5/16$	$1\,33/64$	$37/64$
1	$1\,1/2$	$1\,47/64$	$43/64$
$1\,1/8$	$1\,11/16$	$1\,61/64$	$3/4$
$1\,1/4$	$1\,7/8$	$2\,11/64$	$27/32$
$1\,3/8$	$2\,1/16$	$2\,3/8$	$29/32$
$1\,1/2$	$2\,1/4$	$2\,19/32$	1
$1\,3/4$	$2\,5/8$	$3\,1/32$	$1\,5/32$
2	3	$3\,15/32$	$1\,11/32$

Nominal Dimensions of Heavy Hex Bolts and Heavy Hex Cap Screws

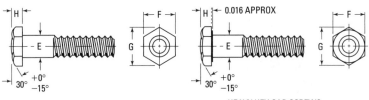

HEAVY HEX BOLTS   HEAVY HEX CAP SCREWS

| | *Heavy Hex Bolts* | | *Heavy Hex Cap Screws* | |
| | *Width* | | *Height* | |
Nominal Size	*Across Flats* **F**	*Across Corners* **G**	*Bolts* **H**	*Screws* **H**
$\frac{1}{2}$	$\frac{7}{8}$	1	$\frac{11}{32}$	$\frac{5}{16}$
$\frac{5}{8}$	$1\frac{1}{16}$	$1\frac{15}{64}$	$\frac{27}{64}$	$\frac{25}{64}$
$\frac{3}{4}$	$1\frac{1}{4}$	$1\frac{7}{16}$	$\frac{1}{2}$	$\frac{15}{32}$
$\frac{7}{8}$	$1\frac{7}{16}$	$1\frac{21}{32}$	$\frac{37}{64}$	$\frac{35}{64}$
1	$1\frac{5}{8}$	$1\frac{7}{8}$	$\frac{43}{64}$	$\frac{39}{64}$
$1\frac{1}{8}$	$1\frac{13}{16}$	$2\frac{3}{32}$	$\frac{3}{4}$	$\frac{11}{16}$
$1\frac{1}{4}$	2	$2\frac{5}{16}$	$\frac{27}{32}$	$\frac{25}{32}$
$1\frac{3}{8}$	$2\frac{3}{16}$	$2\frac{17}{32}$	$\frac{29}{32}$	$\frac{27}{32}$
$1\frac{1}{2}$	$2\frac{3}{8}$	$2\frac{3}{4}$	1	$\frac{15}{16}$
$1\frac{3}{4}$	$2\frac{3}{4}$	$3\frac{11}{64}$	$1\frac{5}{32}$	$1\frac{3}{32}$
2	$3\frac{1}{8}$	$3\frac{39}{64}$	$1\frac{11}{32}$	$1\frac{7}{32}$

Nominal Dimensions of Heavy Hex Structural Bolts

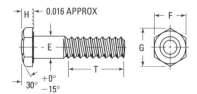

Nominal Size E	Width Across Flats F	Width Across Corners G	Head Height H	Thread Length T
$\frac{1}{2}$	$\frac{7}{8}$	1	$\frac{5}{16}$	1
$\frac{5}{8}$	$1\frac{1}{16}$	$1\frac{15}{64}$	$\frac{25}{64}$	$1\frac{1}{4}$
$\frac{3}{4}$	$1\frac{1}{4}$	$1\frac{7}{16}$	$\frac{15}{32}$	$1\frac{3}{8}$
$\frac{7}{8}$	$1\frac{7}{16}$	$1\frac{21}{32}$	$\frac{35}{64}$	$1\frac{1}{2}$
1	$1\frac{5}{8}$	$1\frac{7}{8}$	$\frac{39}{64}$	$1\frac{3}{4}$
$1\frac{1}{8}$	$1\frac{13}{16}$	$2\frac{3}{32}$	$\frac{11}{16}$	2
$1\frac{1}{4}$	2	$2\frac{5}{16}$	$\frac{25}{32}$	2
$1\frac{3}{8}$	$2\frac{3}{16}$	$2\frac{17}{32}$	$\frac{27}{32}$	$2\frac{1}{4}$
$1\frac{1}{2}$	$2\frac{3}{8}$	$2\frac{3}{4}$	$\frac{5}{16}$	$2\frac{1}{4}$

Nominal Dimensions of Hex Nuts, Hex Thick Nuts, and Hex Jam Nuts

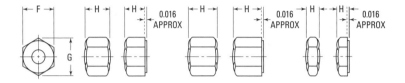

Nominal Size E	Width Across Flats F	Width Across Corners G	Thickness		
			Hex Nuts H	Thick Nuts H	Jam Nuts H
$1/4$	$7/16$	$1/2$	$7/32$	$9/32$	$5/32$
$5/16$	$1/2$	$9/16$	$17/64$	$21/64$	$3/16$
$3/8$	$9/16$	$21/32$	$21/64$	$13/32$	$7/32$
$7/16$	$11/16$	$51/64$	$3/8$	$29/64$	$1/4$
$1/2$	$3/4$	$55/64$	$7/16$	$9/16$	$5/16$
$9/16$	$7/8$	1	$31/64$	$39/64$	$5/16$
$5/8$	$15/16$	$1^3/32$	$35/64$	$23/32$	$3/8$
$3/4$	$1^1/8$	$1^{19}/64$	$41/64$	$13/16$	$27/64$
$7/8$	$1^5/16$	$1^{33}/64$	$3/4$	$29/32$	$31/64$
1	$1^1/2$	$1^{47}/64$	$55/64$	1	$35/64$
$1^1/8$	$1^{11}/16$	$1^{61}/64$	$31/32$	$1^5/32$	$39/64$
$1^1/4$	$1^7/8$	$2^{11}/64$	$1^1/16$	$1^1/4$	$23/32$
$1^3/8$	$2^1/16$	$2^3/8$	$1^{11}/64$	$1^3/8$	$25/32$
$1^1/2$	$2^1/4$	$2^{19}/32$	$1^9/32$	$1^1/2$	$27/32$

Nominal Dimensions of Square-Head Bolts

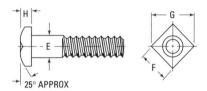

25° APPROX

Nominal Size E	Width Across Flats F	Width Across Corners G	Head Height H
$3/8$	$9/16$	$51/64$	$1/4$
$7/16$	$5/8$	$57/64$	$19/64$
$1/2$	$3/4$	$1\,1/16$	$21/64$
$5/8$	$15/16$	$1\,21/64$	$27/64$
$3/4$	$1\,1/8$	$1\,19/32$	$1/2$
$7/8$	$1\,5/16$	$1\,55/64$	$19/32$
1	$1\,1/2$	$2\,1/8$	$21/32$
$1\,1/8$	$1\,11/16$	$2\,25/64$	$3/4$
$1\,1/4$	$1\,7/8$	$2\,21/32$	$27/32$
$1\,3/8$	$2\,1/16$	$2\,59/64$	$29/32$
$1\,1/2$	$2\,1/4$	$3\,3/16$	1

Nominal Dimensions of Heavy Hex Nuts and Heavy Hex Jam Nuts

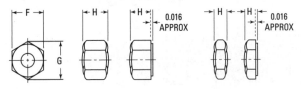

Nominal Size	Width Across Flats F	Width Across Corners G	Thickness	
			Hex Nuts H	Hex Jam Nuts H
$1/4$	$1/2$	$37/64$	$15/64$	$11/64$
$5/16$	$9/16$	$21/32$	$19/64$	$13/64$
$3/8$	$11/16$	$51/64$	$23/64$	$15/64$
$7/16$	$3/4$	$55/64$	$37/64$	$17/64$
$1/2$	$7/8$	$1 1/64$	$31/64$	$19/64$
$9/16$	$15/16$	$1 5/64$	$35/64$	$21/64$
$5/8$	$1 1/16$	$1 7/32$	$39/64$	$23/64$
$3/4$	$1 1/4$	$1 7/16$	$47/64$	$27/64$
$7/8$	$1 7/16$	$1 21/32$	$55/64$	$31/64$
1	$1 5/8$	$1 7/8$	$63/64$	$35/64$
$1 1/8$	$1 13/16$	$2 3/32$	$1 7/64$	$39/64$
$1 1/4$	2	$2 5/16$	$1 7/32$	$23/32$
$1 3/8$	$2 3/16$	$2 17/32$	$1 11/32$	$25/32$
$1 1/2$	$2 3/8$	$2 1/2$	$1 15/32$	$27/32$
$1 5/8$	$2 9/16$	$2 61/64$	$1 19/32$	$29/32$
$1 3/4$	$2 3/4$	$3 11/64$	$1 23/32$	$31/32$
$1 7/8$	$2 15/16$	$3 25/64$	$1 27/32$	$1 1/32$
2	$3 1/8$	$3 39/64$	$1 31/32$	$1 3/32$

Nominal Dimensions of Square Nuts and Heavy Square Nuts

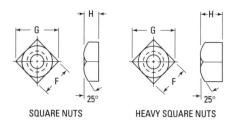

SQUARE NUTS   HEAVY SQUARE NUTS

	Square Nuts		Heavy Square Nuts			
	Width Across Flats		Width Across Corners		Thickness	
Nominal Size	Regular F	Heavy F	Regular G	Heavy G	Regular H	Heavy H
1/4	7/16	1/2	5/8	45/64	7/32	1/4
5/16	9/16	9/16	51/64	51/64	17/64	5/16
3/8	5/8	11/16	57/64	31/32	21/64	3/8
7/16	3/4	3/4	1 1/16	1 1/16	3/8	7/16
1/2	13/16	7/8	1 5/32	1 15/64	7/16	1/2
5/8	1	1 1/16	1 27/64	1 1/2	35/64	5/8
3/4	1 1/8	1 1/4	1 19/32	1 49/64	21/32	3/4
7/8	1 5/16	1 7/16	1 55/64	2 1/32	49/64	7/8
1	1 1/2	1 5/8	2 1/8	2 19/64	7/8	1
1 1/8	1 11/16	1 13/16	2 25/64	2 9/16	1	1 1/8
1 1/4	1 7/8	2	2 21/32	2 53/64	1 3/32	1 1/4
1 3/8	2 1/16	2 3/16	2 59/64	3 3/32	1 13/64	1 3/8
1 1/2	2 1/4	2 3/8	3 3/16	3 23/64	1 5/16	1 1/2

Nominal Dimensions of Lag Screws

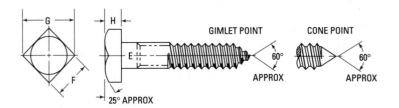

Nominal Size E	Width Across Flats F	Width Across Corners G	Head Height H
	$9/32$	$19/64$	$1/8$
$1/4$	$3/8$	$17/32$	$11/64$
$5/16$	$1/2$	$45/64$	$13/64$
$3/8$	$9/16$	$11/16$	$1/4$
$7/16$	$5/8$	$57/64$	$19/64$
$1/2$	$3/4$	$1 1/16$	$21/64$
$5/8$	$15/16$	$1 21/64$	$27/64$
$3/4$	$1 1/8$	$1 19/32$	$1/2$
$7/8$	$1 5/16$	$1 55/64$	$19/32$
1	$1 1/2$	$2 1/8$	$21/32$
$1 1/8$	$1 11/16$	$2 25/64$	$3/4$
$1 1/4$	$1 7/8$	$2 21/32$	$27/32$

American Standard Machine Screws* (Heads May Be Slotted or Recessed)

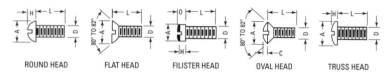

ROUND HEAD   FLAT HEAD   FILISTER HEAD   OVAL HEAD   TRUSS HEAD

Nominal Diameter	Round Head A	H	Flat Head A	Filister Head A	H	O	Oval Head A	C	Truss Head A	H
0	0.113	0.053	0.119	0.096	0.045	0.059	0.119	0.021		
1	0.138	0.061	0.146	0.118	0.053	0.071	0.146	0.025	0.194	0.053
2	0.162	0.069	0.172	0.140	0.062	0.083	0.172	0.029	0.226	0.061
3	0.187	0.078	0.199	0.161	0.070	0.095	0.199	0.033	0.257	0.069
4	0.211	0.086	0.225	0.183	0.079	0.107	0.225	0.037	0.289	0.078
5	0.236	0.095	0.252	0.205	0.088	0.120	0.252	0.041	0.321	0.086
6	0.260	0.103	0.279	0.226	0.096	0.132	0.279	0.045	0.352	0.094
8	0.309	0.120	0.332	0.270	0.113	0.0156	0.0332	0.052	0.384	0.102
10	0.359	0.137	0.385	0.313	0.130	0.180	0.385	0.060	0.448	0.118
12	0.408	0.153	0.438	0.357	0.148	0.205	0.438	0.068	0.511	0.134
¼	0.472	0.175	0.507	0.414	0.170	0.237	0.507	0.079	0.573	0.150
5⁄16	0.590	0.216	0.635	0.518	0.211	0.295	0.635	0.099	0.698	0.183
3⁄8	0.708	0.256	0.762	0.622	0.253	0.355	0.762	0.117	0.823	0.215
7⁄16	0.750	0.328	0.812	0.625	0.265	0.368	0.812	0.122	0.948	0.248
½	0.813	0.355	0.875	0.750	0.297	0.412	0.875	0.131	1.073	0.280
9⁄16	0.938	0.410	1.000	0.812	0.336	0.466	1.000	0.150	1.198	0.312
5⁄8	1.000	0.438	1.125	0.875	0.375	0.521	1.125	0.169	1.323	0.345
¾	1.250	0.547	1.375	1.000	0.441	0.612	1.375	0.206	1.573	0.410

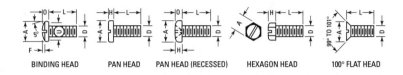

| BINDING HEAD | PAN HEAD | PAN HEAD (RECESSED) | HEXAGON HEAD | 100° FLAT HEAD |

Nominal	*Binding Head*				*Pan Head*			*Hexagon Head*		*100° Flat Head*
Diameter	**A**	**O**	**F**	**U**	**A**	**H**	**O**	**A**	**H**	**A**
2	0.181	0.046	0.018	0.141	0.167	0.053	0.062	0.125	0.050	
3	0.208	0.054	0.022	0.162	0.193	0.060	0.071	0.187	0.055	
4	0.235	0.063	0.025	0.184	0.219	0.068	0.080	0.187	0.060	0.225
5	0.263	0.071	0.029	0.205	0.245	0.075	0.089	0.187	0.070	
6	0.290	0.080	0.032	0.226	0.270	0.082	0.097	0.250	0.080	0.279
8	0.344	0.097	0.039	0.269	0.322	0.096	0.115	0.250	0.110	0.332
10	0.399	0.114	0.045	0.312	0.373	0.110	0.133	0.312	0.120	0.385
12	0.454	0.130	0.052	0.354	0.425	0.125	0.151	0.312	0.155	
1/4	0.513	0.153	0.061	0.410	0.492	0.144	0.175	0.375	0.190	0.507
5/16	0.641	0.193	0.077	0.513	0.615	0.178	0.218	0.500	0.230	0.635
3/8	0.769	0.234	0.094	0.615	0.740	0.212	0.261	0.562	0.295	0.762

*ANSI B18.6—1972. Dimensions given are maximum values, all in inches. Thread length: screws 2 in. long or less, thread entire length; screws over 2 in. long, thread length = 1 3/4 in. Threads are coarse or fine series, class 2. Heads may be slotted or recessed as specified, excepting hexagon form, which is plain or may be slotted if so specified. Slot and recess proportions vary with size of fastener; draw to look well.

American Standard Hexagon Socket,* Slotted Headless,^A and Square-Head^B Setscrews

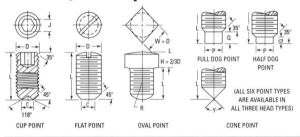

Diameter D	Cup and Flat-Point Diameter C	Oval-Point Radius R	Cone-Point Angle Y		Full and Half Dog Points			Socket Width J
			118° for These Lengths and Shorter	90° for These Lengths and Longer	Length			
					Diameter P	Full Q	Half q	
5	1/16	3/32	1/8	3/16	0.083			1/16
6	0.069	7/64	1/8	3/16	0.092	0.07	0.03	1/16
8	5/64	1/8	3/16	1/4	0.109	0.08	0.04	5/64
10	3/32	9/64	3/16	1/4	0.127	0.09	0.04	3/32
12	7/64	5/32	3/16	1/4	0.144	0.11	0.06	3/32
1/4	1/8	3/16	1/4	5/16	5/32	1/8	1/16	1/8
5/16	11/64	15/64	5/16	3/8	13/64	5/32	5/64	5/32
3/8	13/64	9/32	3/8	7/16	1/4	3/16	3/32	3/16
7/16	15/64	21/64	7/16	1/2	19/64	7/32	7/64	7/32
1/2	9/32	3/8	1/2	9/15	11/32	1/4	1/8	1/4
9/16	5/16	27/64	9/16	5/8	25/64	9/32	9/64	1/4
5/8	23/64	15/32	5/8	3/4	15/32	5/16	5/32	5/16
3/4	7/16	9/16	3/4	7/8	9/16	3/8	3/16	3/8
7/8	33/64	21/32	7/8	1	21/32	7/16	7/32	1/2
1	19/32	3/4	1	1 1/8	3/4	1/2	1/4	9/16
1 1/8	43/64	27/32	1 1/8	1 1/4	27/32	9/16	9/32	9/16
1 1/4	3/4	15/16	1 1/4	1 1/2	15/16	5/8	5/16	5/8
1 3/8	53/64	1 1/32	1 3/8	1 5/8	1 1/32	11/16	11/32	5/8
1 1/2	29/32	1 1/8	1 1/2	1 3/4	1 1/8	3/4	3/8	3/4
1 3/4	1 1/16	1 5/16	1 3/4	2	1 5/16	7/8	7/16	1
2	1 7/32	1 1/2	2	2 1/4	1 1/2	1	1/2	1

*ANSI B18.3—1976. Dimensions are in inches. Threads coarse or fine, class 3A. Length increments: 1/4 in. to 5/8 in. by (1/16 in.); 5/8 in. to 1 in. by (1/8 in.); 1 in. to 4 in. by (1/4 in.); 4 in. to 6 in. by (1/2 in.). Fractions in parentheses show length increments; for example, 5/8 in. to 1 in. by (1/8 in.) includes the lengths 5/8 in., 3/4 in., 7/8 in., and 1 in.

^A ANSI B18.6.2—1972. Threads coarse or fine, class 2A. Slotted headless screws standardized in sizes No. 5 to 3/4 in. only. Slot proportions vary with diameter. Draw to look well.

^B ANSI B18.6.2—1972. Threads coarse, fine, or 8-pitch, class 2A. Square-head setscrews standardized in sizes No. 10 to 1 1/2 in. only.

American Standard Cap Screws[a] (Socket[b] and Slotted Heads[c])

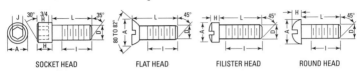

SOCKET HEAD   FLAT HEAD   FILISTER HEAD   ROUND HEAD

Nominal Diameter	Socket Head[d]			Flat Head[e]	Filister Head[e]		Round Head[e]	
	A	H	J	A	A	H	A	H
0	0.096	0.060	0.050					
1	0.118	0.073	0.050					
2	0.140	0.086	1/16					
3	0.161	0.099	5/64					
4	0.183	0.112	5/64					
5	0.205	0.125	3/32					
6	0.226	0.138	3/32					
8	0.270	0.164	1/8					
10	5/16	0.190	5/32					
12	11/32	0.216	5/32					
1/4	3/8	1/4	3/16	1/2	3/8	11/64	7/16	3/16
5/16	7/16	5/16	7/32	5/8	7/16	13/64	9/16	15/64
3/8	9/16	3/8	5/16	3/4	9/16	1/4	5/8	17/64
7/16	5/8	7/16	5/16	13/16	5/8	19/64	3/4	5/16
1/2	3/4	1/2	3/8	7/8	3/4	21/64	13/16	11/32
9/16	13/16	9/16	3/8	1	13/16	3/8	15/16	13/32
5/8	7/8	5/8	1/2	1 1/8	7/8	27/64	1	7/16
3/4	1	3/4	9/16	1 3/8	1	1/2	1 1/4	17/32
7/8	1 1/8	7/8	9/16	1 5/8	1 1/8	19/32		
1	1 5/16	1	5/8	1 7/8	1 5/16	21/32		
1 1/8	1 1/2	1 1/8	3/4					
1 1/4	1 3/4	1 1/4	3/4					
1 3/8	1 7/8	1 3/8	3/4					
1 1/2	2	1 1/2	1					

[a]Dimensions in inches.
[b]ANSI B18.3—1976.
[c]ANSI B18.6.2—1972.
[d]Thread coarse or fine, class 3A. Thread length l: coarse thread, 2D + 1/2 in.; fine thread, 1 1/2D + 1/2 in.
[e]Thread coarse, fine, or 8-pitch, class 2A. Thread length l: 2D + 1/4 in.
Slot proportions vary with size of screw; draw to look well. All body-length increments for screw lengths 1/4 in. to 1 in. = 1/8 in., for screw lengths 1 in. to 4 in. = 1/4 in., for screw lengths 4 in. to 6 in. = 1/2 in.

English to Metric Conversion Table

	Decimals to Millimeters			Fractions to Decimals to Millimeters					
Decimal	mm	Decimal	mm	Fraction	Decimal	mm	Fraction	Decimal	mm
0.001	0.0254	0.200	5.0800	1/64	0.0156	0.3969	17/64	0.2656	6.7469
0.002	0.0508	0.210	5.3340	1/32	0.0312	0.7938	9/32	0.2812	7.1438
0.003	0.0762	0.220	5.5880	3/64	0.0469	1.1906	19/64	0.2969	7.5406
0.004	0.1016	0.230	5.8420						
0.005	0.1270	0.240	6.0960						
0.006	0.1524	0.250	6.3500	1/16	0.0625	1.5875	5/16	0.3125	7.9375
0.007	0.1778	0.260	6.6040						
0.008	0.2032	0.270	6.8580						
0.009	0.2286	0.280	7.1120	5/64	0.0781	1.9844	21/64	0.3281	8.3344
		0.290	7.3660	3/32	0.0938	2.3812	11/32	0.3438	8.7312
				7/64	0.1094	2.7781	23/64	0.3594	9.1281
0.010	0.2540	0.300	7.6200						
0.020	0.5080	0.310	7.8740						
0.030	0.7620	0.320	8.1280	1/8	0.1250	3.1750	3/8	0.3750	9.5250
0.040	1.0160	0.330	8.3820						
0.050	1.2700	0.340	8.6360						
0.060	1.5240	0.350	8.8900	9/64	0.1406	3.5719	25/64	0.3906	9.9219
0.070	1.7780	0.360	9.1440	5/32	0.1562	3.9688	13/32	0.4062	10.3188
0.080	2.0320	0.370	9.3980	11/64	0.1719	4.3656	27/64	0.4219	10.7156
0.090	2.2860	0.380	9.6520						
		0.390	9.9060						
				3/16	0.1875	4.7625	7/16	0.4375	11.1125
0.100	2.5400	0.400	10.1600						
0.110	2.7940	0.410	10.4140						
0.120	3.0480	0.420	10.6680	13/64	0.2031	5.1594	29/64	0.4531	11.5094
0.130	3.3020	0.430	10.9220	7/32	0.2188	5.5562	15/32	0.4688	11.9062
0.140	3.5560	0.440	11.1760	15/64	0.2344	5.9531	31/64	0.4844	12.3031
0.150	3.8100	0.450	11.4300						
0.160	4.0640	0.460	11.6840						
0.170	4.3180	0.470	11.9380	1/4	0.2500	6.3500	1/2	0.5000	12.7000
0.180	4.5720	0.480	12.1920						
0.190	4.8260	0.490	12.4460						
0.500	12.7000	0.750	19.0500	33/64	0.5156	13.0969	49/64	0.7656	19.4469
0.510	12.9540	0.760	19.3040	17/32	0.5312	13.4938	25/32	0.7812	19.8438
0.520	13.2080	0.770	19.5580	35/64	0.5469	13.8906	51/64	0.7969	20.2406
0.530	13.4620	0.780	19.8120						
0.540	13.7160	0.790	20.0660						
0.550	13.9700			9/16	0.5625	14.2875	13/16	0.8125	20.6375
0.560	14.2240								
0.570	14.4780	0.800	20.3200						
0.580	14.7320	0.810	20.5740	37/64	0.5781	14.6844	53/64	0.8281	21.0344

(continued)

Decimals to Millimeters				Fractions to Decimals to Millimeters					
Decimal	mm	Decimal	mm	Fraction	Decimal	mm	Fraction	Decimal	mm
0.590	14.9860	0.820	20.8280	$^{19}/_{32}$	0.5938	15.0812	$^{27}/_{32}$	0.8438	21.4312
		0.830	21.0820	$^{39}/_{64}$	0.6094	15.4781	$^{55}/_{64}$	0.8594	21.8281
		0.840	21.3360						
		0.850	21.5900						
0.600	15.2400	0.860	21.8440	$^{5}/_{8}$	0.6250	15.8750	$^{7}/_{8}$	0.8750	22.2250
0.610	15.4940	0.870	22.0980						
0.620	15.7480	0.880	22.3520						
0.630	16.0020	0.890	22.6060	$^{41}/_{64}$	0.6406	16.2719	$^{57}/_{64}$	0.8906	22.6219
0.640	16.2560			$^{21}/_{32}$	0.6562	16.6688	$^{29}/_{32}$	0.9062	23.0188
0.650	16.5100			$^{43}/_{64}$	0.6719	17.0656	$^{59}/_{64}$	0.9219	23.4156
0.660	16.7640								
0.670	17.0180	0.900	22.8600						
0.680	17.2720	0.910	23.1140	$^{11}/_{16}$	0.6875	17.4625	$^{15}/_{16}$	0.9375	23.8125
0.690	17.5260	0.920	23.3680						
		0.930	23.6220						
		0.940	23.8760	$^{45}/_{64}$	0.7031	17.8594	$^{61}/_{64}$	0.9531	24.2094
		0.950	24.1300	$^{23}/_{32}$	0.7188	18.2562	$^{31}/_{32}$	0.9688	24.6062
0.700	17.7800	0.960	24.3840	$^{47}/_{64}$	0.7344	18.6531	$^{63}/_{64}$	0.9844	25.0031
0.710	18.0340	0.970	24.6380						
0.720	18.2880	0.980	24.8920						
0.730	18.5420	0.990	25.1460	$^{3}/_{4}$	0.7500	19.0500	1	1.0000	25.4000
0.740	18.7960	1.000	25.4000						

Decimal Equivalents, Squares, Cubes, Square and Cube Roots, and Circumferences and Areas of Circles (from 1/64 to 5/8 Inch)

Fraction	Decimal Equivalent	Square	Square Root	Cube	Cube Root	Circle* Circumference	Circle* Area
						0.04909	0.000192
1/32	0.03125	0.0009766	0.176777	0.000030518	0.31498	0.09817	0.000767
3/64	0.046875	0.0021973	0.216506	0.000102997	0.36056	0.14726	0.001726
1/16	0.0625	0.0039063	0.25	0.00024414	0.39685	0.19635	0.003068
5/64	0.078125	0.0061035	0.279508	0.00047684	0.42749	0.24544	0.004794
3/32	0.09375	0.0087891	0.306186	0.00082397	0.45428	0.29452	0.006903
7/64	0.109375	0.0119629	0.330719	0.0013084	0.47823	0.34361	0.009396
1/8	0.125	0.015625	0.353553	0.0019531	0.5	0.39270	0.012272
9/64	0.140625	0.0197754	0.375	0.0027809	0.52002	0.44179	0.015532
5/32	0.15625	0.0244141	0.395285	0.0038147	0.53861	0.49087	0.019175
11/64	0.171875	0.0295410	0.414578	0.0050774	0.55600	0.53996	0.023201
3/16	0.1875	0.0351563	0.433013	0.0065918	0.57236	0.58905	0.027611
13/64	0.203125	0.0412598	0.450694	0.0083809	0.58783	0.63814	0.032405
7/32	0.21875	0.0478516	0.467707	0.010468	0.60254	0.68722	0.037583
15/64	0.234375	0.0549316	0.484123	0.012875	0.61655	0.73631	0.043143
1/4	0.25	0.0625	0.5	0.015625	0.62996	0.78540	0.049087
17/64	0.265625	0.0705566	0.515388	0.018742	0.64282	0.83449	0.055415
9/32	0.28125	0.0791016	0.530330	0.022247	0.65519	0.88357	0.062126
19/64	0.296875	0.0881348	0.544862	0.026165	0.66710	0.93266	0.069221
5/16	0.3125	0.0976562	0.559017	0.030518	0.67860	0.98175	0.076699
21/64	0.328125	0.107666	0.572822	0.035328	0.68973	1.03084	0.084561
11/32	0.34375	0.118164	0.586302	0.040619	0.70051	1.07992	0.092806
23/64	0.359375	0.129150	0.599479	0.046413	0.71097	1.12901	0.101434
3/8	0.375	0.140625	0.612372	0.052734	0.72112	1.17810	0.110445
25/64	0.390625	0.1525879	0.625	0.059605	0.73100	1.22718	0.119842
13/32	0.40625	0.1650391	0.637377	0.067047	0.74062	1.27627	0.129621
27/64	0.421875	0.1779785	0.649519	0.075085	0.75	1.32536	0.139784
7/16	0.4375	0.1914063	0.661438	0.083740	0.75915	1.37445	0.150330

(continued)

Fraction	Decimal Equivalent	Square	Square Root	Cube	Cube Root	Circle*	
						Circumference	Area
$^{29}/_{64}$	0.453125	0.2053223	0.673146	0.093037	0.76808	1.42353	0.161260
$^{15}/_{32}$	0.46875	0.2197266	0.684653	0.102997	0.77681	1.47262	0.172573
$^{31}/_{64}$	0.484375	0.2346191	0.695971	0.113644	0.78535	1.52171	0.184269
$^{1}/_{2}$	0.5	0.25	0.707107	0.125	0.79370	1.57080	0.196350
$^{33}/_{64}$	0.515625	0.265869	0.718070	0.137089	0.80188	1.61988	0.208813
$^{17}/_{32}$	0.53125	0.282227	0.728869	0.149933	0.80990	1.66897	0.221660
$^{35}/_{64}$	0.546875	0.299072	0.739510	0.163555	0.81777	1.71806	0.234891
$^{9}/_{16}$	0.5625	0.316406	0.75	0.177979	0.82548	1.76715	0.248505
$^{37}/_{64}$	0.578125	0.334229	0.760345	0.193226	0.83306	1.81623	0.262502
$^{19}/_{32}$	0.59375	0.352539	0.770552	0.209320	0.84049	1.86532	0.276884
$^{39}/_{64}$	0.609375	0.371338	0.780625	0.226284	0.84780	1.91441	0.291648
$^{5}/_{8}$	0.625	0.390625	0.790569	0.244141	0.85499	1.96350	0.306796

*Fraction represents diameter.

Decimal Equivalents, Squares, Cubes, Square and Cube Roots, and Circumferences and Areas of Circles (from $^{41}/_{64}$ to 1 Inch)

Fraction	Decimal Equivalent	Square	Square Root	Cube	Cube Root	Circle* Circumference	Area
$^{41}/_{64}$	0.640625	0.410400	0.800391	0.262913	0.86205	2.01258	.322328
$^{21}/_{32}$	0.65625	0.430664	0.810093	0.282623	0.86901	2.06167	.338243
$^{43}/_{64}$	0.671875	0.451416	0.819680	0.303295	0.87585	2.11076	.354541
$^{11}/_{16}$	0.6875	0.472656	0.829156	0.324951	0.88259	2.15984	.371223
$^{45}/_{64}$	0.703125	0.494385	0.838525	0.347614	0.88922	2.20893	.388289
$^{23}/_{32}$	0.71875	0.516602	0.847791	0.371307	0.89576	2.25802	.405737
$^{47}/_{64}$	0.734375	0.539307	0.856957	0.396053	0.90221	2.30711	.423570
$^{3}/_{4}$	0.75	0.5625	0.866025	0.421875	0.90856	2.35619	.441786
$^{49}/_{64}$	0.765625	0.586182	0.875	0.448795	0.91483	2.40528	.460386
$^{25}/_{32}$	0.78125	0.610352	0.883883	0.476837	0.92101	2.45437	.479369
$^{51}/_{64}$	0.796875	0.635010	0.892679	0.506023	0.92711	2.50346	.498736
$^{13}/_{16}$	0.8125	0.660156	0.901388	0.536377	0.93313	2.55254	.518486
$^{53}/_{64}$	0.828125	0.685791	0.910014	0.567921	0.93907	2.60163	.538619
$^{27}/_{32}$	0.84375	0.711914	0.918559	0.600677	0.94494	2.65072	.559136
$^{55}/_{64}$	0.859375	0.738525	0.927024	0.634670	0.95074	2.69981	.580036
$^{7}/_{8}$	0.875	0.765625	0.935414	0.669922	0.95647	2.74889	.601320
$^{57}/_{64}$	0.890625	0.793213	0.943729	0.706455	0.96213	2.79798	.622988
$^{29}/_{32}$	0.90625	0.821289	0.951972	0.744293	0.96772	2.84707	.645039
$^{59}/_{64}$	0.921875	0.849854	0.960143	0.783459	0.97325	2.89616	.667473
$^{15}/_{16}$	0.9375	0.878906	0.968246	0.823975	0.97872	2.94524	.690291
$^{61}/_{64}$	0.953125	0.908447	0.976281	0.865864	0.98412	2.99433	.713493
$^{31}/_{32}$	0.96875	0.938477	0.984251	0.909149	0.98947	3.04342	.737078
$^{63}/_{64}$	0.984375	0.968994	0.992157	0.953854	0.99476	3.09251	.761046
1	1	1	1	1	1	3.14159	.785398

*Fraction represents diameter.

Screw Threads and Tap Drill Sizes

NC or ASME Special Machine Screws				NF or ASME Special Machine Screws			
Size of Tap	Threads per Inch	Tap Drill	Body Drill	Size of Tap	Threads per Inch	Tap Drill	Body Drill
1	64	53	48	2	64	50	44
2	56	50	44	3	56	45	39
3	48	47	39	4	48	42	33
4	40	43	33	5	44	37	$1/8$
5	40	38	$1/8$	6	40	33	28
6	32	36	28	8	36	29	19
8	32	29	19	10	32	21	11
10	24	25	11	10*	30	22	11
12	24	16	$7/32$	12	28	14	$7/32$

American Standard Taper Pipe Threads			NF or SAE Standard Screws		
Size of Tap	Threads per Inch	Tap Drill	Size of Tap	Threads per Inch	Tap Drill
$1/8$	27	$11/32$	$1/4$	28	3
$1/4$	18	$7/16$	$5/16$	24	1
$3/8$	18	$19/32$	$3/8$	24	0
$1/2$	14	$23/32$	$7/16$	20	$25/64$
$3/4$	14	$15/16$	$1/2$	20	$25/64$
1	$11 1/2$	$1 5/32$	$9/16$	18	$33/64$
$1 1/4$	$11 1/2$	$1 1/2$	$5/8$	18	$37/64$
$1 1/2$	$11 1/2$	$1 23/32$	$11/16$	16	$5/8$
2	$11 1/2$	$2 3/16$	$3/4$	16	$11/16$
$2 1/2$	8	$2 5/8$	$7/8$	14	$13/16$
3	8	$3 1/4$	1	14	$15/16$
			$1 1/8$	12	$1 3/64$

*ASME only. Tap drills allow approx. 75% full thread.

Number and Letter Sizes of Drills with Decimal Equivalents*

Drill No.	Fractional	Decimal	Drill No.	Fractional	Decimal	Drill No.	Fractional	Decimal
80	—	0.0135	48	—	0.0760	18	—	0.170
79	—	0.0145	—	5/64	0.0781	—	11/64	0.172
—	1/64	0.0156	47	—	0.0785	17	—	0.173
78	—	0.0160	46	—	0.0810	16	—	0.177
77	—	0.0180	45	—	0.0820	15	—	0.180
76	—	0.0200	44	—	0.0860	14	—	0.182
75	—	0.0210	43	—	0.0890	13	—	0.185
74	—	0.0225	42	—	0.0935	—	3/16	0.188
73	—	0.0240	—	3/32	0.0938	12	—	0.189
72	—	0.0250	41	—	0.0960	11	—	0.191
71	—	0.0260	40	—	0.0980	10	—	0.194
70	—	0.0280	39	—	0.0995	9	—	0.196
69	—	0.0292	38	—	0.1015	8	—	0.199
68	—	0.0310	37	—	0.1040	7	—	0.201
—	1/32	0.0313	36	—	0.1065	—	13/64	0.203
67	—	0.0320	—	7/64	0.1094	6	—	0.204
66	—	0.0330	35	—	0.1100	5	—	0.206
65	—	0.0350	34	—	0.1110	4	—	0.209
64	—	0.0360	33	—	0.1130	3	—	0.213
63	—	0.0370	32	—	0.116	—	7/32	0.219
62	—	0.0380	31	—	0.120	2	—	0.221
61	—	0.0390	—	1/8	0.125	1	—	0.228
60	—	0.0400	30	—	0.129	A	—	0.234
59	—	0.0410	29	—	0.136	—	15/64	0.234
58	—	0.0420	—	9/64	0.140	B	—	0.238
57	—	0.0430	28	—	0.141	C	—	0.242
56	—	0.0465	27	—	0.144	D	—	0.246
—	3/64	0.0469	26	—	0.147	—	1/4	0.250
55	—	0.0520	25	—	0.150	E	—	0.250
54	—	0.0550	24	—	0.152	F	—	0.257
53	—	0.0595	23	—	0.154	G	—	0.261
—	1/16	0.0625	—	5/32	0.156	—	17/64	0.266
52	—	0.0635	22	—	0.157	H	—	0.266
51	—	0.0670	21	—	0.159	I	—	0.272
50	—	0.0700	20	—	0.161	J	—	0.277
49	—	0.0730	19	—	0.166	—	9/32	0.281

(continued)

Drill No.	Fractional	Decimal	Drill No.	Fractional	Decimal	Drill No.	Fractional	Decimal
K	—	0.281	Y	—	0.404	—	45/64	0.703
L	—	0.290	—	13/32	0.406	—	23/32	0.719
M	—	0.295	Z	—	0.413	—	47/64	0.734
—	19/64	0.297	—	27/64	0.422	—	3/4	0.750
N	—	0.302	—	7/16	0.438	—	49/64	0.766
—	5/16	0.313	—	29/64	0.453	—	25/32	0.781
O	—	0.316	—	15/32	0.469	—	51/64	0.797
P	—	0.323	—	31/64	0.484	—	13/16	0.813
—	21/64	0.328	—	1/2	0.500	—	53/64	0.828
Q	—	0.332	—	33/64	0.516	—	27/32	0.844
R	—	0.339	—	17/32	0.531	—	55/64	0.859
—	11/32	0.344	—	35/64	0.547	—	7/8	0.875
S	—	0.348	—	9/16	0.562	—	57/64	0.891
T	—	0.358	—	37/64	0.578	—	29/32	0.906
—	23/64	0.359	—	19/32	0.594	—	59/64	0.922
U	—	0.368	—	39/64	0.609	—	15/16	0.938
—	3/8	0.375	—	5/8	0.625	—	61/64	0.953
V	—	0.377	—	41/64	0.641	—	31/32	0.969
W	—	0.386	—	21/32	0.656	—	63/64	0.984
—	25/64	0.391	—	43/64	0.672	—	1	1.000
X	—	0.397	—	11/16	0.688			

*Sizes start with No. 80 and go up to 1 inch. This table is useful for quickly determining the nearest drill size for any decimal, for root diameters, body drills, etc.

Index

milling machines *(continued)*
attachments and accessories,
    263–265
basic construction and
    classification, 233–239
bed-type, 238–239
cleaning, 242, 265
coolants for, 261, 265
defined, 233, 239
dividing heads, 267–278
duplex, 239, 240
feed, 257–261, 265
hobbing machines, 32
knee-and-column, 233–236, 240
locating the cutter, 254–255
micrometer table attachment,
    263, 264
mounting the cutter, 250–253
mounting the work on,
    242–246, 265
oiling, 242, 265
operation of, 233, 239–240
plain, 235–237, 240
plain horizontal, 236–237
planer type, 32
selecting the cutter, 246–247
sharpening the cutter, 247–250
speed, 255–257, 265
spindle, 235
spline mills, 238, 240
thread mills, 238, 240
types of, 236–239
universal, 237, 238
universal dividing head, 262,
    263
uses of, 233, 240
vertical, 237, 239
vertical milling attachments, 265
minimum allowance, 173
minor diameter, 175
mortising, 71, 81
motor reverse control
    (for tapping), 66, 67
multiple indexing, 298, 299
multiple-spindle lathe, 132
multiple toolholder, 228
multiple-tool lathe, 132

multi-spindle drilling machines,
    45, 48

N
negative compounding, 291–22
neutral space, 175
nicked cutter, 246
nominal swing, 107
nonferrous saws, 18–19, 20
nonprecision grinding, 353, 363
number and letter sizes of drills
    with decimal equivalents,
    422–423
number of threads, 175

O
odd-geared lathe, 177
offhand grinding. *See*
    nonprecision grinding
offset slide, turret lathe, 212
oil-hole twist drill, 57
open-side planer, 325, 338
outside diameter, 175

P
parallels, 75, 77, 314, 315
pedestal grinder, 353
pitch, 175
pitch diameter, 175
plain cutter, 246
plain dividing head, 268–269,
    270, 274, 278
plain horizontal milling machine,
    236
plain indexing, 284–289, 299
    calculation of on the standard
        driving head, 286–289
    and compound indexing
        combined, 294
    defined, 284
    dividing-head sector, use of,
        284–286
    gearing arrangement, 284, 299
plain milling machines, 235–237,
    240
plain slide, turret lathe, 211
plain vise, 314, 315